NOT AFRAID OF LIFE

not afraid of

my journey so far

BRISTOL PALIN
with Nancy French

wm
WILLIAM MORROW
An Imprint of HarperCollins*Publishers*

HarperCollins books may be purchased for educational, business, or sales promotional use. For information please write: Special Markets Department, HarperCollins Publishers, 10 East 53rd Street, New York, NY 10022.

FIRST EDITION

Designed by Jamie Lynn Kerner

Library of Congress Cataloging-in-Publication Data has been applied for.

ISBN 978-0-06-208937-3

11 12 13 14 15   OV/RRD   10 9 8 7 6 5 4 3 2 1

To all you underdogs

It is not the critic who counts; not the man who points out how the strong man stumbles or where the doer of deeds could have done better. The credit belongs to the man who is actually in the arena, whose face is marred by dust and sweat and blood, who strives valiantly, who errs and comes up short again and again, because there is no effort without error and shortcomings, but who knows the great enthusiasms, the great devotions, who spends himself for a worthy cause; who, at the best, knows, in the end, the triumph of high achievement, and who, at the worst, if he fails, at least he fails while daring greatly, so that his place shall never be with those cold and timid souls who knew neither victory nor defeat.

—Theodore Roosevelt

Contents

NOT AFRAID OF LIFE

Introduction

I lied to my mother.

"We're going to go stay the night at Ema's house," I nonchalantly said as my friend and I headed toward the front door. Mom was busy paying bills and didn't really look up from her work. There was only one week left of school, and the weather was warming up in the Matanuska-Susitna Borough.

"Okay," she responded, not suspecting a thing. "Do you need me to drop you off?"

"No," I said. "Her mom's in a hurry, but she's going to pick us up at the end of the driveway."

"Have fun," she said casually and waved good-bye.

That deception would affect my life in ways a teenager could not comprehend. It changed my relationship with my parents, my boyfriend, and even God. It would eventually cause me public embarrassment on an international scale and cause many sleepless nights.

But I didn't know that at the time.

On that day, my friend and I believed we were getting away with a harmless high school lie. Usually I have a very sensitive conscience, even to the point that I can't leave a store without fixing a messed-up clothes rack. I think, I'm going to fix these or some overworked employee is going to have to do it later and she's probably already done it a million times today. But on that morning, my conscience wasn't even really pricked. Apparently, the excitement of seeing Levi outweighed any anxiety I felt about lying to my mom. So, we toted our bags down to the end of the long gravel driveway, jumped in his red pickup truck, and left without any sort of guilt.

As we drove away from my house, I drove away from the ease of childhood and smack into the middle of the weird complexities of serious relationships ideally reserved for later in life. We drove for about an hour, deep into the Point MacKenzie area that is sparsely populated with almost perpetual sunlight during the summer months. We loved it because of its amazing wildlife and natural beauty. My friend and I couldn't imagine a more exciting night than hanging with our friends in such a setting. In the back of Levi's truck were tents, sleeping bags, firearms for protection against wildlife, and lots of alcohol.

I never drank—in fact, I knew nothing about anything bad really . . . especially the differences between vodka, beer, and whiskey. I didn't know that the girly flavored wine coolers were just as likely to get you drunk as the hard stuff, even though they went down so smoothly. And I definitely had no idea what "tolerance" was or how to pace your drinking to make sure you don't do things you'll regret. All I knew was that I was with my ruggedly handsome boyfriend who loved me—and we were getting away with a late-night camping trip without anyone ever finding out.

In fact, *no one* has ever heard this part of my story. By now, most of America knows me as Bristol Palin, the teenager who got pregnant right before her mother was asked to run for vice president on the GOP ticket with Senator John McCain. But what no one really knows is my story—the true story—of deception and disappointment that began the night I lied to my mother and went camping in Point MacKenzie.

We got there around six o'clock. Levi and his friends immediately built a fire and put up the tents by the lake. The tent my friend and I brought was blue, and had just enough space for both of us to squeeze in for a good night's rest. However, I didn't end up sleeping in that tent.

The wine coolers tasted sweet, and I slowly surrendered to their woozy charms. I felt young and carefree, and Levi kept replacing my empty bottles from his large stash. The more I drank, the better the crisp night air felt. But unbeknownst to me, I was about to hit a wall—that awful wall—that takes you past a comfortable level of libation—the happy buzz—into the dark abyss of drunkenness.

I remember sitting in one of those folding camping chairs, laughing with friends by the fire.

What I don't remember is what transpired between the moment when I was sitting there by the fire talking and the moment I awakened the next morning with something obviously askew. Mosquitoes were buzzing around my ears and my head throbbed like someone was using it as a drum. Levi's empty sleeping bag was right beside mine, and I could hear him outside the tent laughing as he and his friends packed up the camp.

I fumbled around for my phone and found it in a pile of clothes on the side of the tent.

Get over here

I texted my friend.

Within seconds, she unzipped the tent and poked her head in.

"Are you okay?"

"What happened?" I whispered.

"You don't know?"

In movies, losing your virginity is a big deal . . . a candlelit experience with romantic music, roses, and declarations of true love. I'd said—repeatedly—that I'd save sex for the first time on my wedding night. Brought up in a Christian household, I was determined to do things in the right order. Dating first, followed by an engagement, a beautiful wedding night, a romantic honeymoon, and—only then—the figurative baby carriage immortalized in the kids' "first comes love, then comes marriage" taunt.

That's why the next sentence that came out of my friend's mouth hit me like a punch in the stomach.

"You *definitely* had sex with Levi."

Suddenly, I wondered why it was called "losing your virginity," because it felt more like it had been stolen.

"No, we didn't," I insisted.

My friend didn't argue with me, because I could tell by the evidence in the tent that all of my plans, my promises, and my moral standards had disappeared in one awful night in a series of bad decisions.

And Levi wasn't even there to help me process—or even confirm—my greatly feared suspicions. Instead of waking up in his arms (which happens in all the movies right before the girl walks around the apartment in the guy's buttoned-up business shirt), I awakened in a cold tent alone as he talked with his friends on the other side of the canvas.

I didn't realize this was a sign of things to come. But I did know one thing.

I was going to marry Levi. I had to now.

As a person who was raised to believe that sex should be reserved for marriage, I wasn't sure how to handle the fact that I royally betrayed my parents' and my moral code. I wish I'd confessed my sin right then to God, accepted the full forgiveness of a heavenly father who loved me, and never spoken to Levi again. Instead I tried to salvage this situation. I tried to fix it.

After all, my own wonderful dad had lived a pretty tough life compared to mine before my parents were married. Similarly, Levi could turn into the godly man I knew he could be. And I naively assumed I was just the girl who could show him the way.

As I sit here in my parents' home in Wasilla several years later, I realize how stupid that must seem to readers who saw parts of my drama playing out in real time. You may have followed my mom's meteoric rise to fame and political prominence, only to have the gnat named Levi Johnston constantly spreading false accusations against our family. You may have seen an on-again-off-again engagement in the media and wondered why I'd put up with someone who cheated on me about as frequently as he sharpened his hockey skates or as frequently as he filled up his old truck.

The truth is that my teenage brain believed I could pick up the pieces of my shattered moral code and glue it back together. The secret that I'd lost my virginity became the unfortunate guiding compass throughout the rest of high school.

Why am I telling you—and my family for the first time—this personal information?

Statistics show that my story, sadly, is not a unique one. Seven in ten teens have had sex by the time they turn nineteen, frequently in spite of their best intentions and moral beliefs. Since most people don't marry until their mid-twenties, young adults are at increased risk of unwanted pregnancy and sexually transmitted diseases for nearly a decade.

Just as a person afflicted with lung cancer might feel compelled to warn others about making the same mistake, I feel it is necessary—and helpful—to be candid about how I ended up making some pretty foolish decisions.

I know honestly dealing with problems publicly can make a huge difference in people's lives. In fact, recently, I got a message from a stranger who messaged me through my Facebook page.

> Hey Bristol,
> I have no idea if you remember me, but I met you last year right when I found out I was pregnant. I honestly believe that if it weren't for you I would've had an abortion. You have become somewhat of a hero to me. Hearing your story has encouraged me to fight and try to create a good life for my son. Just wanted to send some thanks and stay strong!

When I read that, I realized people are deeply impacted by other people's decisions, and somehow, in God's amazing providence, I played a role in saving a baby's life.

Though I'm not an emotional person, I cried as I read her note and vowed to be more candid about my experiences.

Maybe my story was told on a national stage because that teenager in Texas needed a little inspiration to encourage her to keep her baby. If that's what it took to save that baby's life from an abortion clinic, I'd do it fifteen more times. All of that heartache and public humiliation would've been worth it.

But I have a suspicion that my story will impact many more people to reconsider how they date, when they have sex, and what they do if they've already made some pretty terrible decisions. I'm not a role model and definitely not a preacher. I'm just a normal girl who couldn't hide her problems from a gawking world and learned a few lessons along the way.

Sometimes, not being afraid of life in all of its imperfections is the first step toward a better future. That's true of my own life. But before I tell you about the behind-the-scenes scoop on my mom's vice presidential bid, that two-week engagement announced in *Us Weekly,* and why on earth I agreed to be on *Dancing with the Stars,* you have to understand where my story began.

In a little place called Wasilla.

Chapter One

Where It All Began

A few years ago, if I'd told you I was from Wasilla, you would've looked at me blankly with a polite smile, trying to search your brain for any sort of connection. You probably haven't been to Alaska, even though it's America's largest state, and you probably hadn't heard of my hometown before John McCain chose my mom to be his running mate for vice president. Shocked reporters (undoubtedly prepared for a more conventional choice) scrambled to outdoor shops to buy heavy coats and then to the airport to fly to our small town of ten thousand people. Some weren't even aware of how to pronounce the name.

These reporters gave the whole world a crash course on my town, however. Some said it was a beautiful town, surrounded by mountain ranges, eagles, and lakes that turned into gigantic playgrounds for snowmachines and cross-country skiing for months out of the year. They described the amazing wildlife and the clean cool air. But others described it as an ugly city of rednecks who shopped at half-abandoned strip malls.

I just know it—and love it—as home. It's surrounded by mountain ranges, so no matter where you go, you feel protected . . . cradled almost . . . by these majestic formations. We have a lot of cloudy days, which hide the peaks, but it's amazing when the sun comes out. People smile, have a quicker step, and enjoy the sun more because it's relatively rare.

I was raised here because my mom's parents moved to Alaska five years after it became a state, where teachers could make twice the salary ($6,000 per year!). Grandpa was ready for change. He'd had a rather harsh background, which pushed him when he was in high school out of the house and onto friends' couches. He was attracted to Grandma's conventional, stable family life and was determined to create the good, loving family he grew up craving.

They moved with two-year-old child Chuck Jr., one-year-old Heather, and three-month-old baby Sarah.

I wonder what Grandpa would've thought if someone told him that his infant (a girl at that!) would someday be the governor of America's forty-ninth state?

They rented a tiny wooden house in Skagway, the site of the 1898 Klondike gold rush, where at one time 40,000 people lived. By the time my mom's family arrived, the population had dwindled to 650. Make that 656, because after the Heath family of five moved in, they added one more to the family. Aunt Molly was the first member of my mom's side of the family to be born in Alaska.

They fished, hunted, and hiked in that amazing untamed wilderness. Mom's dad taught school, coached ball, seasonally tended bar in tourist areas, and worked on the Alaskan railroad. After saving money for a new and bigger house, in the early 1970s, they moved from Anchorage to Wasilla in the Matanuska-Susitna (Mat-Su) Valley.

Though my grandpa wasn't much for "organized religion," he "forced" the kids to go to church with Grandma. It was at a Bible camp one summer when Mom gave her life to Jesus. Later, she and her siblings were all baptized together in the cold waters near Big Lake.

This was when Mom started to pray a lot about big things and little things. One day she started to pray about finding the right guy to date. Because Wasilla was such a small town, the boys at school felt like brothers to her instead of possible boyfriends. So she asked God to bring in someone new.

She wasn't overly concerned about boys. In fact, her dad warned her about getting too obsessed with guys too soon. Once she had written a guy's name on her hand. When her dad noticed it, he gave her some advice.

"You have a choice between boys and sports," he said. "You're at an age when I start losing my good athletes because they start liking boys. You can't have both."

Mom played softball and volleyball, ran cross-country and track, and was an excellent point guard, too. In her high school championship game, she hit the game-winning free throws while playing with a stress fracture in her ankle!

It turns out that she *could* have sports and boys—at least if it was the right boy. And the funny thing is, the same dad who shooed off her crush was the one who first introduced the idea of Todd Palin into Mom's head.

Grandpa had gone to school to set up his classroom when he

noticed my dad practicing basketball. He was a native Alaskan—part Yup'ik Eskimo—who'd moved into town from Dillingham and was going to play basketball his senior year. He was talented on the basketball court, hardworking as a commercial fisherman, and smart enough with his money to have lots of toys—a Mustang, a truck, and two snowmachines. When Mom finally laid eyes on him, she literally said, "Thank you, God," because she knew immediately he'd answered her prayers.

Their story had at least a few echoes of Grandpa meeting Grandma, such as Dad being drawn to Mom's big, stable family. These stories were ingrained in me as a kid and helped define my view of love and marriage, and even my sense of identity.

I idealized both my parents' and my grandparents' marriage—Mom and Dad have been married twenty-two years, and Grandpa and Grandma have been married fifty years. Without even realizing it, their wonderful stories shaped my own outlook on men and dating. So when I met my own hard-living man from a troubled family, it didn't faze me as much as perhaps it should've.

But I'm getting ahead of myself.

After Dad and Mom got married, they formed a very good life for themselves.

In the early 1970s, Dad purchased a limited-entry permit from his grandpa in Dillingham, which has the best salmon runs in the world. Just like taxicab medallions in New York City, there are only a certain number of permits out there. And Dad got the best spots to catch salmon in Bristol Bay because they're farther north than the other areas. He always got to take the first shot on the fish coming south, which meant he'd be able to bring more fish in. This job—though lucrative—was seasonal. The salmon runs last only four weeks per year, and he needed to stretch that money throughout the year as he worked other jobs.

After Mom and Dad married in 1988, however, my father landed a job in the North Slope oil fields. Being a "sloper" might not mean anything to you in the lower forty-eight, but around Wasilla it means good money, health insurance, and a challenging job that never gets boring. Alaska's North Slope oil and gas industry is not for everyone. But for people who aren't afraid of long hours or harsh, remote environments, these jobs are highly valued. When my dad got a job working for $14 per hour as a production operator at Prudhoe Bay, my mother was thrilled and maybe a little apprehensive. Many marriages don't survive the sloper schedule and the separation. However, Mom was working two jobs (and Dad was working two), so the idea of the better-paying Slope job was too appealing to pass up. So my parents placed his career—and their marriage—in God's hands.

This was beginning to become a recurring theme in their lives.

Being a sloper meant that every week Dad would drive to Anchorage. There, he'd get on a 737 jet to land in the unfortunately named area of Deadhorse, where he'd then get on a shuttle to get to Prudhoe Bay. He slept in a dormitory-style camp building with other workers and ate at a cafeteria.

During the week he was working, he'd have long days (usually twelve hours), which resulted in lots of overtime. His facility separated water from crude oil and gas before sending the oil down the Trans-Alaska Pipeline and the water back into the ground. He loved the job in spite of the fact that it would sometimes be seventy below zero up there.

In spite of their time apart, they were excited about his new job and about how it would help create a nice future for them as a newly married couple. But, in 1989, their lives—and family—expanded, when Mom gave birth to her first son, Track.

Yes, Track.

His arrival began the long and complicated explanations about how my parents named him—and all of their future children. If you're wondering, Track is so named because he was born during spring track season and my folks loved sports. Seventeen months later, they named me Bristol, but told everyone different reasons for my name. Apparently, Dad grabbed the birth certificate and wrote in my name before Mom could get to it. He told everyone I was named after Bristol Bay—where he had fished since he was a kid—but Mom told everyone I was named after Bristol, Connecticut, home of ESPN, where she had hoped one day to be a sportscaster. Again, I'll have to explain this over and over for the rest of my life, but I like my name because it's very Alaska oriented (or Connecticut oriented, depending on who you believe). I was born on October 18, 1990, Alaska Day, and showcased my native roots with dark hair and eyes.

We lived in a nice three-bedroom house built on the far western boundary of our town. It was located right in the center of downtown, on Wasilla Lake, directly across from the highway. As a toddler, I began showing the same personality that would follow me throughout life. I was a complete perfectionist. My mom says I even potty trained myself at fourteen months, a shocking fact I only now appreciate as a mother. I also pretended to speak fluent Spanish, figuring my brother couldn't understand Spanish anyway, so it was a pretty good trick. Plus, I had an imaginary friend I called "Dudda."

But mostly, I had a very nurturing personality and loved helping babysit all of my cousins and younger siblings. And speaking of siblings, they kept coming. Willow was born when I was four, during the salmon run. That meant that Dad was in Dillingham when Mom went into labor, and he missed her arrival into this world by a few hours! Though Mom and Dad were disappointed, making a living in Alaska sometimes requires a lot of travel, hard choices, and time apart.

Willow's name came from a small community that began in 1897 when miners discovered gold on Willow Creek. Also, she was named after Willow Bay, one of my mother's favorite sports reporters. Because we were four years apart, we were each other's best friends and worst enemies, depending on the day.

While Mom took care of small kids, she filled in as a weekend sports anchor in Anchorage, freelanced for the local paper, and did other odd jobs. In 1992, she ran for city council and was elected to two terms. She helped develop our town's infrastructure, focused on making the politicians fiscally responsible, and ensured that the citizens knew what was going on in the government.

Of course, this was when the direction of her life was set into motion, but we didn't know that then. She gave herself to the responsibilities completely, learned what concerns the citizens of Wasilla had, memorized all the lines in the budget, and took great care not to spend the taxpayers' money too casually. While she fulfilled her duties in "Seat E," she still was simply Mom. Once, she even breast-fed Willow while taping a radio segment on local politics!

I was too young to really be aware of the fact that Mom was "in politics." We just knew she was working for something she believed in.

As Mom learned all the ins and outs of city government however, she started having sharp differences of opinion with the local mayor. He actually wanted to *force* some of the outer areas of the Mat-Su Borough to become a part of the city, to broaden the tax base and become more prominent . . . even though those outside areas didn't *want* to be governed by a city government. Mom has always held the same positions that revolved around small government and maximizing individual freedom, and she believed Wasilla would be better off if the mayor's vision of the future was never realized.

While my mom was off making a difference in our town politically, we had some great bonding time with Dad. From the moment I was little, I followed in my dad's footsteps—or his sled tracks?—and rode on a snowmachine. Yes, in Alaska, we call those motorized snow vehicles "snowmachines," though I understand people in the lower forty-eight call them snowmobiles. I think the difference might have to do with the fact that Alaskans consider these sleds as more of a necessity, and not just arctic toys. Many people who don't have cars have snowmachines, and all kinds of people use them for daily commuting to and from the office. Plus, "snowmobile" just sounds weird. (Really!)

Arctic Cat is the manufacturer of my favorite snowmachines . . . and I'm not just saying that because they sponsor my dad in the Iron Dog! They make smaller snowmachines for kids (called "kitty cats," get it?) that limit the speed and allow little three- to five-year-olds to have their own fun on the snow. I had a little 120, which is what my son, Tripp, has now, and my brother and I rode all over the lake. Sometimes Track would even race in "kitty cat" races, in which racers go in a circle around cones. I loved to watch Track do that, and once or twice I participated in my own races. I never won! But even though we learned how to ride snowmachines early on, we still had mishaps. Once my cousin Payton and I were riding outside Dad's Polaris store. He leaned out the door and told us to knock it off, advice we promptly ignored. That's when Payton rode right into Dad's big old green monster truck, denting it with his helmet!

Another time when I was older, we went on a family ride out to a cabin. We rode all in a row—like ducks—and I was the last one in the row. At first, I was having so much fun . . . looking through the goggles at the big snow-covered trees and mountains. There were no cars, buildings, or other people milling around. But as I watched my family zip through the trails ahead of me, I started getting nervous. What if I had a breakdown, what if I got snagged by some branches? They'd never know it! I'd be lunch for some bear!

After worrying for several miles, finally the inevitable happened. I *did* get hung up . . . barely.

"I got stuck," I yelled when they finally stopped after realizing I wasn't following them. Normally, I wasn't a "drama queen," but I'd gotten a little more fearful with every mile. Finally, when they rode up to me, I threw off my goggles and helmet and yelled, "And I almost died!"

But in spite of our mishaps and dramatics, snowmachining has always been a part of our lives and my childhood. (Piper started riding hers out to the cabin—through trails without cell reception, between trees, over frozen lakes, up hills, and even through an open creek—when she was six . . . and the ride is eleven miles!)

We didn't love snowmachines just because they're fun. We also loved them because we wanted to be like Dad. Even though he maintained his jobs, he was in some ways "Mr. Mom" around the house. While Mom was tackling her new job, he kept up the domestic duties he'd begun while she was on the city council. In fact, he's as good at braiding hair as he is at rebuilding a snowmachine in fifty-mile-per-hour winds. When Mom was away on political trips, he would always help us get ready for school. I remember very distinctly that he put my hair up in three cute little ponytails on the first day of preschool. Dad didn't really know it looked kind of silly, but he had done it simply because that's what I wanted. When my mom picked me up and saw those three pony-tails—one on the right side of my head, one on the left, and one in the back—she just laughed. Not only was it fun to see him look through our closets, shirt by shirt, he let us wear whatever we wanted . . . including hats. I loved to wear them, but not the cute kind that some parents put on their children for nice photos. No, I wore cowgirl hats, French berets, and ball caps to school. Though I looked absolutely ridiculous, Dad didn't care. He also let me go to preschool wearing a flower girl dress I'd worn in my aunt and uncle's wedding. Sounds cute until you realize I wore it with black tights and my pink and white Nike Air Force 1's.

And speaking of that flower girl dress . . . I got a lot of wear out of that thing.

When I was four years old, Mom came home and asked, "Hey, Bristol, do you want to be in a pageant?" She was friends with the pageant director, who'd talked her into entering me in it.

At only four, I didn't have enough sense to say no.

Let's just say we weren't quite prepared as we journeyed to an enormous theater in Anchorage on pageant day. It was absolutely, packed! Mom put me in my old flower girl dress, dropped me off with a kiss, and went to sit in the audience with Grandma and Grandpa. When I walked backstage, it was into a cloud of hair spray and makeup. Crazy pageant moms were running around with huge bags of clothespins and duct tape (for emergencies) as they smudged lipstick and eyeliner on their kids' little round faces. I sat down and it dawned on me, more with every passing minute, that I was underprepared for this event. I counted the minutes until it was my turn to take a turn on the stage. Though Mom had participated in a pageant before, she had no idea that other moms would take a kids' pageant so seriously. She sat happily in the audience and smiled as she saw me walk hesitantly onto the stage.

There was just one problem. We were supposed to speak.

Now, I'm not the shyest person in the world, but I was not ready to make my public-speaking debut right then and there.

When Mom realized I was supposed to talk in front of everyone, her face fell. She hadn't prepared me for that. Her eyes were big with worry as I approached the kid-height microphone.

They didn't expect me to deliver the Gettysburg Address. All I had to do was walk up and introduce myself—basic stuff. I was supposed to say my name, hometown, age, and where I attended preschool. But when I got to the mic, I opened my mouth, looked out into the crowd, and froze. The only thing that came out was, "I'm . . . B-B-Bristol . . ."

Then I was quickly overcome with embarrassment and walked off the stage.

I walked out of the theater with the smallest trophy possible, and was the only contestant who didn't get flowers.

The lady who'd given me the trophy smiled—with perfectly applied lipstick—and said, "Thanks for participating."

The plastic gold(ish) trophy was meant to smooth over my terrible performance. But even though I was four, I knew.

I'd bombed.

Even though I wasn't the pageant type, my aunt Molly, who had only a son at the time, loved to doll me up. Once I went to stay with her for the weekend so I could play with my cousin Payton, who's two years younger than I am. Aunt Molly had so much fun spoiling me. She dressed me in sweet dresses and headbands . . . she curled my hair. That one visit, though, she got carried away, threw me in the car seat, and took me to get my ears pierced! I was only two and my mom was a little surprised when Molly dropped me off at our house with pink rhinestone studs in.

"What did you do to my daughter?" she exclaimed.

Aunt Molly, who's now a pediatric dental hygienist, just laughed. Several years later, she had a girl of her own. McKinley just turned ten . . . and she *still* doesn't have her ears pierced!

"Hey, McKinley, wanna go get those ears pierced? I'll take ya!" my mom frequently says jokingly at family functions.

Mom hasn't convinced her yet.

I had a different kind of fashion mishap at my other aunt's house. My parents were out of town, so we were being watched by my aunt Heather and uncle Kurt. I was only about seven or eight years old and had really long beautiful dark hair that I was always known for. It hung all the way to my waist.

"Will you cut my hair? I want to look like Lauden," I said, handing Uncle Kurt a pair of scissors.

(I was always in awe of how beautiful my cousin was—still is!—and I wanted to look just like her!)

Uncle Kurt didn't hesitate. He just put all of that gorgeous hair into a ponytail—and cut the whole thing off. I loved my new short cut until I realized that it was permanent!

My childhood was full of many funny moments like those with our family.

On Sundays we went to church, which—of course—was less than exciting for a little kid. Mom would let me put my head in her lap, and she'd tuck strands of my hair behind my ear, over and over. I'd listen to the preacher talk about Jesus and forgiveness and love, but eventually his voice would seem to grow distant and I'd succumb to sleep right there in the pew.

Sermon sleep is the best sleep ever.

I always looked forward to communion, because after church they let me go through all of the aisles and pick up the little plastic cups.

As I got older, I'd also volunteer to help in the nursery. Even though there were always a bunch of adults in there, I'd be the one who wanted to fuss over the babies and rock them. I just loved babies—real ones!—like most girls love toys or dolls. I'd even take Willow's old car seat, stick a doll in it, cover it with a blanket, and walk around at parks pretending I had a real baby inside. (I thought I tricked a few people, but they may have just been polite.) On the morning of my ninth birthday, I even crawled into bed with Mom and demanded—quite rudely—"If you don't have a baby, you better go rent me one!"

With every year that I grew older, my mom grew more po-litically prominent. When I was seven, Mom decided to run for mayor, and like everything in the Palin household, it was a family affair. She decided on the theme "Positive-ly Palin," and I helped her select the rather unusual combination of pink and green for her signs. (No one had ever used those colors!) We put them all over town. I say "we," but I think I spent more time in the little red wagon with Track while Mom, Dad, and her friends nailed them up. Still, it was a family endeavor, and we were thrilled when she won. And by a pretty good margin!

Soon afterward, in 1999, she was elected by other mayors in Alaska to serve as president of the Conference of Mayors. To out-siders who'd never heard of my mom's name until she burst out onto the national stage during the 2008 presidential campaign, it seemed that her rise to political fame was sudden and abrupt. Un-doubtedly, it was. But as her child, I saw it as a natural, gradual progression, and I never thought anything of being "Sarah Palin's daughter."

In Alaska, I was known just as commonly as "Todd Palin's daughter." He's a legend around here. Not only is he an amazing hair braider, he's a commercial fisherman, had a great job on the North Slope, and was part owner of Valley Polaris. His company sold ATVs, watercraft, and snowmachines and fixed them there in the shop. Willow loved hanging out there so much that Mom said she was raised at the Polaris shop on Dad's hip. Even when she was young, she was a "motorhead."

Dad has won the Iron Dog competition four times and placed second four times . . . an impressive feat since it's the world's longest snowmachine race through the most remote and rugged terrain in Alaska. Of the six hundred or so teams that have started the race since it first began, less than half have finished. Why? Temperatures frequently fall to fifty degrees below zero—not even factoring in the wind—which means Dad wears duct tape on his face for protection. The two-thousand-mile race takes six days, and the trail carries the racers over tree stumps, cliffs, large mounds of earth, the frozen Bering Sea, and other rivers so destructive to snowmachines that when the machines cross the finish line, they have basically been almost completely rebuilt along the way. The drivers don't fare too much better. Broken bones are expected, and many riders just quit because their machines get fried or they tire of the relentless, unimaginable cold. But not my dad. When Mom was governor, people called him the "First Dude," but he was known for being so tough he could withstand wipeouts at one hundred miles per hour and the mechanical breakdowns that would make normal men give up.

My friends may have thought Mom was cool, but they thought Dad was Superman.

Life with my parents was wonderful, though I never really considered their jobs as anything unusual. Maybe Mom had more late-night phone calls and Dad was gone more than some other jobs required. However, our family of five was a fun and great way to grow up. One morning, Mom nudged me from the couch where Willow and I were watching television.

"Come on," she said. "Go put your shoes on, so you guys can come to my doctor's appointment with me."

We piled in the black four-door Bronco, thinking we were running some of our normal errands. We were too young to realize we were in an ob/gyn office. Even if we had noticed the sign on the door, we were too young to know what that even meant.

The nurse came in, put a wand on Mom's belly, and an image popped up on a screen.

"Can you tell the girls what they are looking at?" She beamed at the nurse.

"This," she said very sweetly, "is your new little . . . sister."

It was the first time Mom knew the gender of the baby. And the first time we knew we were getting another sibling.

Finally! I'd wanted a baby in the house to take care of for so long! When my new sister arrived on March 19—exactly on the date she was due—they named her Piper Indi Grace, after the Piper plane my dad flies, the idea of "independence," and the grace of God. Mom let me decide how to spell "Indi," whether with a *y* or an *i*. I chose the *i* because it seemed like it'd be a lot more fun writing than the old boring *y*. Plus, I didn't want people to think of the Indy 500, the big manly Polaris snowmachine that my dad sold at his shop.

Mom's friends gave her a baby shower at the Grouse Ridge Shooting Range because they knew Mom loved the Second Amendment and because they'd shot clay pigeons at the range when Mom was pregnant. Since she was the mayor, it was a big baby shower. The theme, of course, centered around airplanes. The cake was in the shape of a Piper plane, and there were blankets with Pipers on them. I loved being there with Mom, celebrating my new sister's arrival. I also loved that there were three babies under the age of three months there. My uncle's wife had a baby at the end of December, my aunt Molly had a baby at the end of January, and Mom had Piper in the middle of March. I was in heaven!

Though I loved taking care of younger siblings and cousins, that didn't mean that I couldn't hang with the boys. I went to Iditarod Elementary School, and I took swimming lessons, ran track and cross-country, and played soccer. I could run so fast I could beat the boys in our mile run in P.E. class! Plus, I was proud of myself for being elected treasurer of our school, though I don't remember ever handling real money. I also played trumpet in fifth grade, though I was not good! I only chose the trumpet because it had three buttons, but I still complained about having to practice. Mom and Dad would not let me quit, though. I signed up for it, and had to deal with the consequences.

I was always so close to Aunt Molly that she told me about both of her next pregnancies before she told the rest of the family . . . much to my mother's chagrin! But her husband, my uncle Mike, was no prize.

Though sometimes Uncle Mike was charming and fun, he was known around town for embellishing facts and telling outright lies. He was a big burly state trooper . . . six foot four and 250 pounds.

He was very intimidating and always teased Payton by accusing him of being weak.

I saw it firsthand. One day in 2003, my cousin Payton and I were sitting downstairs at their house with him, while Aunt Molly gave her daughter a bath upstairs. That day started out like any other day, but it would become a part of my consciousness, and—sadly—a part of the national political conversation years later when my mom ran for vice president. And the whole controversy started with this stupid question:

"Hey, Payton, do you want to get shot with a Taser gun?"

Yes, a state trooper—an adult—asked that question of a kid.

I could tell Payton was unsure about it, but he didn't want to be accused by his stepdad—for the millionth time—of being a "wuss."

"Okay," he said, staring at the Taser gun that his stepdad pulled out of his holster. "I guess."

Uncle Mike prepped the Taser, and Payton started getting more and more nervous. He didn't let on. I never really thought Uncle Mike would actually go through with it. Perhaps he was just testing him.

But I was wrong. I was standing at the top of the stairs when Uncle Mike took the Taser and shot my cousin. Payton instantly fell back as intense signals were sent through his nervous system. His muscles constricted. In a bit, the pain lessened and he shook his head, as if to get rid of that feeling.

As Payton was recovering, Uncle Mike looked at me and saw that I was crying.

"Bristol, you're next."

I was not about to let what I just saw happen to my cousin happen to me, even if he was an adult. Even if he was my uncle.

"Aunt Molly! Uncle Mike just shot Payton with a Taser!"

"No, he didn't," she said. It's not that she didn't think I was trustworthy; she just couldn't imagine that the man she married would do such a thing to an innocent little kid.

When it finally dawned on her that I was telling the truth—and Payton confirmed it—she was livid.

We weren't the only people who had trouble with Mike. In fact, there were so many citizen complaints against him that one day a state trooper detective asked to come to our home and interview us about what we'd observed about his activities.

"Of course," we said. Cooperating with a detective is not something you think twice about.

What was revealed about Mike's actions as a state trooper was shocking, including citizens claiming to have watched him chug beer *in his patrol car,* claims that he later denied; and one of his fellow police officers confessed that he witnessed him illegally killing an animal on a hunting trip, another claim that he also denied.

Oddly, the detective didn't seem too interested in my story. In fact, he turned off his tape recorder during our interview and lectured me about how Taser guns aren't lethal. I told him about my Internet research on the dangers of using a Taser on a little kid, including Taser-related fatalities, but he just laughed.

"The Internet is full of lies," he said.

My interview only seemed to help Mike and his fellow union members (including the detective himself!) make it sound like I was just whining about a cop. I sure wasn't whining. I was just truthfully answering the questions. This is when I really began to wrangle with the ideas of justice and fairness, and when I learned some people just don't want to be bothered by the truth. The incident also showed that my mom and dad cared so much about family that they were ready to stick their necks out, even at great cost to themselves.

It was a regrettable incident that stuck out because I had an otherwise peaceful childhood. Our home was right in the middle of town, so that meant our friends were always over. Dad paved us a basketball court area and put up a hoop. We putzed around on the lake on Jet-Skis and boats. On the Fourth of July, friends would come over and watch the fireworks over the lake and go swimming. Our house was known as the place to hang out after school or practice, and Mom kept the house full of cookies. We'd also go out to Papa Jim's cabin in Crosswinds (Papa Jim is my paternal grandfather) and shoot BB guns and go berry picking for pies we'd make for dinner at his cabin. My parents worked hard to make our family a good one.

It might seem unconventional that Mom was the mayor and Dad was both a commercial fisherman and worked in the North Slope . . . and it was. But through love and a lot of effort, they made the challenging jobs and demanding schedules work for our family. I grew up knowing how to bait a hook, I grew up knowing how to shoot a basketball, and I grew up knowing that I was loved. In many ways, my childhood was very similar to my mom's . . . full of fishing, hunting, camping, family, church, and friends.

Then, in seventh grade, I met a guy named Levi.

Chapter Two

First Impressions

I n seventh grade, my locker was right beside Levi Johnston's. Of course I noticed he was handsome, and I was happy when we'd run into each other between classes. However, I hadn't thought much about him until an English teacher gave all of us an assignment: write a letter to ourselves for a time capsule to be opened at the end of high school. The assignment asked us to list our goals, our dreams, and our interests. I remember thinking and thinking about my list because I wanted it to be just right. After all, it seemed like a big deal to put something into a "time capsule." I didn't realize it only meant the teacher would file it under "Class of 2009" in some dusty old file room. At any rate, my letter had some lofty goals, like wanting to own fifty pairs of jeans, having my own pig, wanting to go to a Lakers game, and meeting President Bush. But at the very end, after I'd sat there thinking about who I might like, my mind went back to that cocky guy whose locker was next to mine. I'd already written "I have a crush on . . ." So as the teacher told us to turn in our papers, I hurriedly scribbled "Levi Johnston."

That's how it all started.

That year, we lingered a little too long at our lockers, so long that we'd almost miss class. Which was silly, since we had math, English, social studies, and science together anyway. We saw plenty of each other.

Levi was known for stirring up trouble in class, especially when we had a substitute teacher. But one day he picked the wrong substitute teacher to challenge—my grandpa. An amazing thing about school was that Grandpa frequently substituted for us in science. And he was great! His house is full of antlers, snake skins, monkey skulls, duck bones, owl pelts, jars of dissected animals, and porcupine quills to show the class. My friends loved it, because when they saw Grandpa, they knew they'd learn more interesting things than just the "parts of a cell." Plus, he never assigned homework because he believed kids should get their schoolwork done during school hours.

On this particular day, Grandpa was there teaching us about petrified wood, when Levi started his usual antics. He wasn't loud and obnoxious, he was quiet and obnoxious. He threw spitballs at someone, kicked a kid under the table, and even punched some poor kid. When Grandpa kicked him out of the class and sent him to the office, it was the first sign of how well Levi would get along with my family.

He didn't make a good first impression on my mom, either. Later that year, his competition hockey team played my older brother Track's team and won. Afterward, Levi came up to me and gloated, "I told you I'd beat Track's team." Mom, who was standing right there, was not impressed with this big guy who had a talent for trash-talking but nothing else.

"What's that guy's problem, and why is he gloating to you about your brother's loss?"

I assured her that he was just kidding, but I could tell Mom was not fond of him. I quickly ran and hid in the car, so he wouldn't run his mouth again.

Though I loved my mom and grandpa and respected their opinions, Levi's bad behavior wasn't a deal breaker for me. I found him exciting. Like many women throughout history, I went for the "bad boy" who didn't care about authority. After all, he was my opposite. I was a rule follower, a teacher's pet, a straight-A student. I didn't even cuss, and when people used bad words around me, I'd correct them. "Come on," I'd say a little self-righteously. "Don't use that kind of language."

Levi was my first crush, though nothing "official" was ever established between us to give us the designation of "boyfriend and girlfriend." However, it felt more "official" when we had a substitute teacher (not Grandpa!), and Levi made the class memorable for me in a way different from his usual antics. Instead of using this opportunity to pull someone's hair or trip someone, in the middle of class he reached under the table and grabbed my hand. It was the first time I'd ever held hands with a boy, and my heart raced.

Throughout the year, we passed notes to each other all the time, under our teachers' noses. Once a note landed on my desk and I carefully unfolded the paper. I gasped when I read what he'd written.

Will u be my gurl?

No, it didn't have boxes to check—yes, no, or maybe. It was just the one sentence, all alone, in a rather lame attempt to make sure I was "his."

I didn't think about it too long. I scribbled hurriedly:

No! You're supposed to ask me in person.

Levi could barely get through a week without getting in trouble at Teeland Middle School. He had a sister named Mercede (like the car, but without the *s*) whom we called "Sadie." She wasn't a model student either, and his mom eventually transferred both of them to a different school. This was a good indication that things at his home were really not good. His mom was always letting her kids do whatever they wanted. (She was later arrested for selling OxyContin in the local Target parking lot and served time in jail.)

When Levi finally left my school, a lot of drama left with him, and I sank all of my energy into sports. Athletics shaped my life as much as the state in which I grew up . . . just as they did for my grandparents and both of my parents. I played basketball, ran track, was the captain of my volleyball team, wrestled, and then I added another, unexpected sport to my roster.

It happened because I idolized my older brother, Track. When he made the offhand comment that football practice was harder than basketball, well, I had to show him he was wrong.

"It is not," I protested. "And I'll prove it to you."

"How can you do that?" he asked, familiar with the stubborn streak that made me unable to back down from any fight. "You've never played football."

He was right, of course. After all, girls my size didn't play football, but I never let anything keep me from a challenge. "I'll join the team then."

And so I did, much to the bemused pleasure of my brother.

I was one of the smallest receivers the Houston Hawks had, and we lost every game. However, I went to every practice, kept up with the boys, and never complained. At the time, my mom was training for a marathon, so she'd drive me to practice and run to get a couple of hours of training in while we were out on the field.

Normally, my brother was very protective. Though he would later threaten to beat up anyone who messed with me, this "protective streak" disappeared when it came to our bet. I could tell the guys on my team were a little hesitant to tackle Track Palin's sister. But he'd laugh and say, "Go ahead. Hit her harder!"

After weeks of practice, it was time for the actual games to begin. One Saturday, I was running a play when someone came at me. I got tackled so hard that I lay on the ground and cried . . . right in front of my brother and his friends in the stands. I saw my dad, who was standing on the sidelines, wince. Lying there on the field, I'd never felt more embarrassed in my life. I was practically sobbing, which wasn't pretty since I'd gotten all the wind knocked out of me. It was the only time I cried.

I want to quit, I remember thinking. But quitting wasn't an option. At least not for me. I went back onto the field and played. We didn't score any points at all, until the last game. The one touchdown we scored wasn't enough to win a game, but we were thrilled! That season, I gained respect from my teammates and coaches because I'd toughed it out without whining. When that football season finally ended, so did my short-lived football career.

And I survived it, Track Palin. You have to admit that now.

Though I was the only girl to play football, my friends and I stuck together in our other athletic activities.

My best friends were Jenna, Ema, and Sammy. Jenna's dad was a sloper like mine, so she was a hard worker and was focused on getting good grades, too. It was nice, because we shared all the same classes. Her family belonged to the Mormon church, and they were always involved in church activities. Ema went to the same Bible church as my family did, but she was a very curious girl, always pushing the limits.

In a word, we were jocks. In Wasilla, there wasn't the typical "cheerleader" stereotype like there is in the other states. The popular girls were the athletic ones, and we played everything our town offered.

On a whim, my friend Sammy and I decided to join the wrestling team. (Not as uncommon up here as a girl being on the football team.) This entailed learning hundreds of moves, both offensively and defensively and from the top and bottom. After much practice, when the referee blew the whistle at the beginning of the match, and my opponent began circling and looking for an opening or to clinch my arms, I was ready. We worked so hard during that season, and we were in the best shape of our lives! Dad never came to my meets, because he did not want to see his daughter wrestling. Football was one thing, but this? It was too much for him. The worst part about the whole experience was that we had to wear purple "singlets" for our meets. They looked so ridiculous that Sammy and I wore them to volleyball practice to get a few laughs. We looked stupid, but it's fun to be silly when you have your friends around you.

The girls and I also played on the basketball team together. In eighth grade, I was a shooting guard for Teeland Middle School's team—the Titans—and proudly wore the number 20. (My mother was number 22 all through her high school basketball days.) Our coach, John Brown, was a talented guy with light blond hair and three daughters. He was also our math teacher. This meant that we basketball players tried to get away with everything in class.

Jenna and I would go up to him in the middle of prealgebra and ask, "Can we go do the basketball laundry?" We got out of a lot of boring math that way, but we still managed to make all As. In fact, all of our middle school teachers made learning fun. For example, in eighth grade, we took a multimedia class where we shot our own videos. Jenna and I took on the challenging task of making a video called "How to Make a Grilled Cheese Sandwich." (This, sadly, is still about the extent of my cooking skills.)

Even though we'd get away with mischief in class, Coach Brown was all business on the court. He taught us tons of plays that we'd later use in high school. The girls on the team had a friendship that went beyond just learning various plays. On game days, Jenna, Ema, Sammy, and I all wore matching basketball shoes that coordinated with our purple and silver uniforms. Plus, we'd always do our hair in the same way, all pulled back on the sides or in French braids. At lunch, we'd always sit together and talk about our strategies and upcoming games. And even after games, Coach Brown would make us write reflection pages that made us think back and judge what we'd done well and what we'd done poorly.

Once, we were coming back from basketball regionals at Kenai, when the car Jenna's mom was driving hit black ice and spun out of control. She had a pretty bad accident, with Jenna and some other kids in the car. It just so happened that my grandma and grandpa were driving right behind them, so they stopped, called an ambulance, got the kids blankets to keep warm, and helped everyone stay calm while they waited. Other than a lot of stitches, everyone was fine. It just goes to show how small and tight-knit my town is—we all come to each other's aid in times of need.

Since we enjoyed sports so much, we also went to watch Wasilla High School games. In small towns, there is rarely anything going on more exciting than these well-attended community events. One night, Jenna, Ema, Sammy, and I bundled up and went to a football game. I can't remember if we won or lost, but I do remember that Sadie and a girl named Lanesia were there. I'd gone to school with Lanesia back at Iditarod Elementary School, and we never really got along.

That night, she and Sadie were not happy with us.

We figured this out by picking up on the subtle clues they were giving us. Like when Sadie yelled "I found them!" as they darted around the parking lot. And then Lanesia yelled, "Let's kick their ass!" Okay, maybe not so subtle.

Now, I'd like to say that my friends and I turned around and put a stop to that nonsense. After all, there were four of us and two of them. We probably could've taken them. But the truth is, we were terrified. We'd never fought before, and had never even seen a fight before. So we ran. We ran between the rows of cars, we ducked behind cars. Finally, we saw a parked truck belonging to someone we knew. We tried the handle and realized—thankfully!—it was unlocked. That's where we hid, laughing about what was happening, our hearts racing, as they looked for us until finally giving up.

It was just one of those weird middle school dramas that Jenna, Ema, Sammy, and I lived through together. But we also had political adventures. By 2002, Mom had served two terms as mayor of Wasilla and started thinking about running for a higher office. It seemed like the administrative position of lieutenant governor was a good fit for Mom, because she could put to work the skills she learned as CEO of our city, in the position of mayor. It might sound like her decision to run would've really affected my life, but it didn't. Because she didn't like asking people for money, she didn't have a huge campaign. So when she ran for lieutenant governor, it was a low-key affair that included a lot of fun but not a great deal of pressure. In other words, it didn't have much of an impact on my life.

Every weekend, my friends and I would pile into my mom's four-wheel-drive Bronco and drive around the state to help her at rallies and parades. She was one of six candidates running, and she didn't really have a chance. Dad was working full-time on the North Slope, had just sold the Polaris dealership, had started building a new house on Lake Lucille, and was preparing for the Iron Dog. Mom was coaching youth basketball, had just had Piper, and was still the full-time mayor. Honestly, Mom had just bitten off more than she could chew. (Later, Grandma would tell her, "You can't have it all, all at the same time.") My main memory from that time was that Mom had gotten an awful haircut during the campaign that made her bangs stick straight out. We'd walk up to her with our hand sticking out from our forehead, making fun of the way her hair stuck out. She wasn't fazed by it, though. She just pinned her bangs down and pretended they were *supposed* to look that way. Otherwise, the campaign was only really significant because her showing in the election led to her appointment as chairman of the Alaska Oil and Gas Conservation Commission, then as chairman of the nation's Interstate Oil and Gas Compact Commission. This allowed Mom to live in Wasilla and drive to Anchorage for her job. It actually simplified our lives, because she was no longer the mayor, and she didn't have to constantly be on the phone listening to people who wanted a fire hydrant added to their street or to get her to shut up some barking dog in their neighborhood.

Many late nights, Jenna, Ema, Sammy, and I would have slumber parties and talk about cute boys—we laughed about Sadie and Lanesia's threats, talked about all the things we got to do on the campaign trail, and dreamed about going to high school. *Dreamed* is probably not the right word. Perhaps I should've said *dreaded* being the new kids in the school. Though we were terrified about stepping into the halls for the first time, we knew we'd be able to make it . . . together.

Then, something happened in 2005. After an evening volleyball game, I could suddenly tell that something wasn't right around the house. When I came into their room to say good night, I noticed Mom and Dad cut short their conversation and smiled at me. The next morning, they talked in low voices. I knew something big was going on, but I wasn't a good enough eavesdropper to figure it out.

It was already dark when my mom picked me up from school a couple of days later. In the winter, Alaskan students go to school in the dark and drive home in the dark. At some point during the day the sun might come out momentarily, but we were inside and never saw it. I remember that Dad was on the Iron Dog trail that particular day. Because this race lasts several days, it was an exciting time of life when we'd wonder where he was on the trail, if he was in the lead, and whether he was safe.

When I got in the car and shut the door, my mom's voice was more serious than normal.

"I have something to tell you, but you have to promise not to talk about it at school."

Was there something wrong? Was Dad hurt on the trail? My mind reeled with the possibilities.

"Uncle Mike is having an affair."

The news sank into me like a knife, and my mind was filled with concern over how my aunt Molly and cousins were going to deal with this. In fact, who would want to have an affair with someone who thinks it's perfectly reasonable to Taser a young kid?

I didn't know anyone on my mom's side of the family who had ever been through a messy divorce.

"With Jenna's mom," Mom added. Jenna's mom was a good friend of all of ours.

That ten-minute drive home felt like death.

This wasn't supposed to happen. Jenna came from one of those "good families." I never would've guessed her mother was capable of splitting apart my aunt's family so easily. I was always at their house—even spent school nights there—and my aunt Molly and Jenna's mom were good friends! The affair became public when Jenna's mom told my mom while they were watching us play a volleyball game that season. That's when things unraveled. Jenna's mom was kicked out of their Latter-Day Saints church, Aunt Molly began divorce proceedings, and my cousins prepared to live life without a dad.

The next day when I went to school, I didn't dare mention it. Though I was hanging out with my friends just as much, the secret lingered between us like a ghost. It haunted me. It haunted Jenna. We didn't talk about it. It tore us apart.

It only became a topic of conversation among us when Mike (I no longer called him uncle) showed up at volleyball games with Jenna's mom. It was so awkward, but we tried not to think about the new, weird dynamics in the stands. We tried to focus on setting the volleyball, spiking it over the net, and having good clean fun.

That's before we realized it just wasn't possible anymore. When everyone found out at school, people started choosing sides. Well before we'd ever heard of Team Jolie or Team Aniston, my friends at Teeland unofficially broke into two groups. Those who supported my family, and those who supported Jenna's. I hated the drama.

It was also the first time—but not the last—that I realized how someone's private sexual sin could rock everyone around him or her.

When, in that same year, our teacher assigned us a project to do a report on a famous poem, I chose the John Donne poem from which the line "No man is an island" is taken, because the phrase was something my mom always said to us.

I remember getting up in front of the class and clearing my throat before I read it.

"No man is an island entire of itself; every man is a piece of the continent, a part of the main," I read. The class, I'm sure, wasn't paying much attention to my little speech, as I finished it up, "And therefore never send to know for whom the bell tolls; it tolls for thee."

When I finished reading, the teacher urged me to go on with my explanation of the poem, but I was hesitant because Jenna was in the class. I was afraid she'd be able to tell the entire report was about Mike.

"This poem speaks to me about how people are not as isolated from each other as it might seem. One man's actions can—and will—affect everyone around them in unimaginable ways."

Mike's behavior upset a lot of people's lives, and I hated him for all the drama he caused my family.

I couldn't have predicted I'd soon be causing my family a different kind of drama.

Chapter Three

Losing It

Hockey in Alaska is like basketball in Indiana or football in Tennessee. It's the thing all kids do, and hockey families incur great expense so their kids can play. Sometimes entire teams take trips to other cities for tournaments and games. Since Mom was a "hockey mom," I was forever tagging along behind her, helping her keep the stats, providing the guys with Gatorade, and helping with any other things that came up. In fact, I missed the first day of high school because of a hockey trip to Boston. Track's team—which included Levi, a guy named Ben, and other friends—packed up and went south to Massachusetts. Though it was a fun trip, it put me at a disadvantage when it came to getting into the groove of a new (and bigger!) school.

That's how the second day of school was my first. Track and I drove in his Bronco to Wasilla High School, music blaring, with the base turned up. As the base of the music thumped, so did my heart. How would I know where to go, what if I couldn't find my classes, and where would I sit at lunch? Track was making fun of me on the way, but I couldn't calm my nerves—I was about to become one of the hundreds of confused-looking freshmen.

As soon as I walked in the door, my heart raced. I was trying to find my first class, when—through the crowd of complete strangers—I saw my friend.

"Over here!" I yelled, thankful to see a friendly face. But when she approached, I noticed we were wearing the exact same shirt—a button-up cardigan we'd bought on a shopping trip together. Okay, so I realize now it's not that big of a deal, but for a freshman during her first hour of high school, it was disastrous; I wanted to crawl into my locker and hide. At least we'd bought different colors.

When I did find my first class, I settled into a seat and got my notebook out to jot down any information I'd need for the class. After the class began, Levi walked in late. This second day was his "first day," too, since he'd been on the hockey team's trip.

"Hey, babe," he said as he slid into his chair.

Though I'd never chase or pursue him, I was happy to have a friend next to me. At that time, that's all we were . . . though we started flirting from that first class on. As the days passed, I watched in mild amusement—and some concern—as it became clear he wasn't fitting in well at high school. He came to hockey practice drunk, he put Ex-Lax in some kid's brownie, he got sent to the office. Apparently, Levi only enrolled in school so he could qualify to play hockey. As soon as the season was over, he'd drop out. (It became a pattern. For later semesters, he enrolled in a homeschool curriculum for just enough credits to play. After hockey season, he'd drop out again.) Because he knew he wasn't going to be there for long, he had no reason whatsoever to behave in school. He was a one-man wrecking crew, and he only lasted a few weeks at Wasilla High School.

I was astonished at how little Levi cared about what teachers thought of him, how little he studied, and how he lived on his own terms. After he dropped out, however, it was out of sight, out of mind. The excitement—and challenges—of a new school crowded out any lingering thoughts of rekindling that old seventh-grade flame. I'd sometimes see him at the house with Track and his other teammates, but otherwise I didn't pay much attention to him and did normal high school things.

I still had the drama from the affair of my soon-to-be former uncle to deal with. One night at the beginning of the school year, my friends and I went to the high school football game and were hanging around talking. Mike was an assistant coach at the high school, which made it awkward when I'd run into him.

That night, however, we were walking by him, and—in a great demonstration of both his immaturity and his feelings for his former niece—he called me a "f—ing b—ch!" Though he mumbled it, every one of my friends heard it.

As he walked off, my friends said, "Really? Did that really happen? A trooper just called you that?"

All of the drama surrounding my uncle was hard for me to survive without my naive notions about truth and justice being stripped away. Even though I was so young, I suddenly became calloused toward people in authority. It also changed the way I looked at men and fidelity. And, not least of which, it scared me. After a state trooper told me he would "bring me down" and "get you all," it was enough to cause me to lie awake at night.

When he called me a b—ch that night, it seemed to bring it all back.

My friends and I told the teachers what he'd called me, thinking it would at least get him a slap on the wrist for speaking to a student that way. However, he stayed on as coach, and I just tried to avoid him.

It was just another one of those irritants that kind of stole some of the carefree joy out of high school.

During my freshman year, Mom and Dad gave me a "purity ring," a little silver band with a diamond on it. A few of my friends in Wasilla at the time wore these types of rings, and I was happy when I received one, too. I didn't take a chastity pledge or participate in any sort of ceremony. Mine was a purity ring and was just a simple symbol of what I already knew: I was going to remain pure throughout my high school years.

It didn't seem like such a stretch.

In middle school, I was voted "Most Likely to Succeed," and that same personality carried through to high school. I was even more organized, more dedicated to hard work, and more of a teacher's pet than ever before. Yes, I was that girl. The one who people try not to cuss around. So the purity ring fit right in with my good-girl personality.

Like most kids, my freshman year was a challenge. Though I had good guy friends, the gang from middle school had largely broken up. The tension caused by Jenna's mom and my uncle having an affair made everything more difficult, so Jenna, Ema, Sammy, and I all went our separate ways. I could barely remember what it was like to have late-night chats at slumber parties when we'd talk about high school. Though we'd all been nervous about entering a new environment, we also had different perspectives about ninth grade. Some of my friends were curious about guys and knew that high school meant one thing: losing their virginity. (One proudly wore a T-shirt that read FRESHMAN FIFTEEN. Instead of referring to potential weight gain, though, the slogan was about sex; the shirt had the names of fifteen boys written—and crossed out—underneath it.) I was taking a different approach from some of them, however, and told my girlfriends that I was waiting until marriage.

Of course, my friends and I assumed we would stay close in high school and go through the "coming-of-age" moments together. We had no idea that life would intervene and splinter our relationships. I watched from a distance as my friends lived their separate lives. I played basketball, but it wasn't as much fun now that everyone was on different levels of teams—some were on junior varsity, others on varsity, and so forth. To make matters worse, remember Lanesia? She was now a teammate of mine. Though she wasn't chasing me around the parking lot threatening to kill me, she was always causing drama on the team.

I turned fifteen the day my mother announced she was running for governor of the state of Alaska. Over the summer, Mom had been thinking about whether she should challenge Governor Murkowski, who had already served one term. Many people had called her to complain about how the government had gotten out of control, how they were sick of "politics as usual," and how the oil companies weren't drilling and were robbing hardworking Alaskans of job opportunities. Dad supported Mom, like he has always done. So, on my birthday, Mom announced from our living room that she was running for the office.

Okay, she didn't choose that date because I was turning fifteen. She chose to announce on October 18, 2005, because it was Alaska Day, the anniversary of the day when the United States bought the Territory of Alaska from the Russians in 1867. Kids got out of school, state employees got off work, businesses closed for the day, parades were thrown, and my mom kicked off her gubernatorial run in our living room.

That meant that suddenly our weekends became very full. Every Saturday, we'd load up in Mom's Jetta and head out on long road trips. When we arrived at our destination—we made name tags, blew up balloons, and stole a few Murkowski signs off people's lawns. (Sorry, Mom! Sorry, Governor Murkowski!) We were—as always—working on a limited budget, so we'd frequently spend more money on gas than we'd make at the fund-raiser. That's because Alaska is so gigantic that some of our in-state trips would take seven hours, then we'd turn around and come back to save money on hotels. But we did get to see most of our beautiful state. Frequently, we'd see bear, moose, buffalo, and sometimes wolves walking across the roads or grazing nearby. Some of the roads were built over permanently frozen ground. When the hot pavement was poured, it melted the ground, which caused the ground to "heave" from side to side as well as front to back. This meant that some roads were like roller coasters, with dips sometimes a foot deep! They could wreak havoc on your car . . . and stomach, after too much ice cream.

In spite of all these low-budget road trips, Mom had learned from her lieutenant governor loss to do things right this time. Her two themes were simple and catchy: "New Energy for Alaska" and

"Take a Stand." Plus, she set up a nice campaign office in Anchorage on Fifth Street. One weekend, we loaded the car with paintbrushes, rollers, and several gallons of red paint and painted the Alaska flag on her headquarters wall. As we stood back and admired our work, everything seemed so much more official. Like always, her endeavor was a family affair. Kris Perry, one of her dear friends—and a fellow soccer mom—became instrumental as they figured out how to win the gubernatorial race. Even my great-grandmother on my dad's side got involved in the campaign. Mom called her a "one-woman Eskimo whistle-stop tour" as she went around Dillingham and told the elders about "Todd's wife."

Our fall was consumed with starting that adventure, but right after spring break in 2006 I ran into Levi.

"Come see me play in Fairbanks," he said.

"I'm already going," I told him, trying to hide the fact that I was thrilled at his invitation. That hockey-mom thing my mom always talked about? It wasn't a campaign tactic. She really was the manager of Track's hockey team, which means she went to his games and cheered the young players, kept score, compiled stats, organized transportation, and put bandages on game-induced wounds. Though Fairbanks was a seven-hour drive from our hometown, I happily endured the road trip there. I watched from the stands as Levi played.

A few days after that game, he invited me to the movies. Because we were so young, he had his dad pick me up and drop us off. After seeing a silly movie that made us both laugh, he leaned in to kiss me.

Afterward, he looked me straight in the eye and said, "I just had to do that."

"Why?" I asked, a little skeptical even though I was a freshman. "So you can tell your friends?"

"No," he purred. "So I can sleep tonight."

Okay, so maybe I wasn't skeptical enough, because that line—that cheesy line—melted my heart. The next day I was still intoxicated with new love, when I noticed he was texting someone.

"What's going on?" I asked. "Is that a girl you're texting?"

"Um, yeah," he said. "She was asking me a question about something."

"Really? About what?"

"She's just a friend," he said. "Ask my mom or sister. They'll tell you."

This became a pattern. At first, I'd call to check out his story, but his family always backed him up. It made me feel like I was being petty and small-minded.

Of course, he was texting a girl—because he was with that girl when I wasn't around.

After so much public unfaithfulness, it almost seems strange to mention such a small incident. But it's significant because it was the initial lie snowball that started an avalanche.

Levi would frequently lie to me about other things, too. For example, he told me he had a new cobalt blue truck—a 2007 Chevy Silverado extended cab, with rims, and a double pipe exhaust. Not only did he have no cobalt blue truck, it was before he had a truck at all. Though he later got a series of red trucks, none matched his disturbingly specific description. He also would act like he had lots of money, which is odd since I didn't even really care about that. He'd say he got twelve fish when he didn't catch any. It's weird. He'd lie when the truth would do just fine.

I ignored his lies, mesmerized by what I thought was love, and the year went on well for me at Wasilla High School. With one week left, everyone was looking forward to the wonderful Alaskan summer.

That's when I did the thing that drove me crazy about Levi: I told a lie.

"We're going to go stay the night at Ema's house," I nonchalantly said as my friend and I headed toward the front door and to Point MacKenzie.

Of course, I thought I was headed into an evening of harmless high school fun. But really, I was headed into the deep quicksand of sexual sin, during a night that I barely remembered. The next morning when I woke up, I didn't know what had happened until I spoke to my friend, who confirmed my deepest fears.

On the drive home from Point MacKenzie, she and I sat in the backseat while Levi chatted with his friend riding in the passenger seat. (Guys, take note: Do not put your girlfriend in the backseat so that you can talk hunting with your friend.) His only acknowledgment of our life-changing moment was the occasional knowing glance or wink in the rearview mirror.

He remembers, I thought . . . and dread crept all over me. I tried not to vomit.

When we got back to the house, I was devastated as I tried to figure out what had just happened. When we got a chance to be alone, I pulled him aside and began a very uncomfortable conversation.

"You knew I didn't want to have sex until I was married!" I whispered. "How could you?"

He never really gave me a good answer, but he did apologize. "I thought you wanted to," he said. "I thought you changed your mind."

He also claimed to have been drunk, which I believed.

Okay, there are several times in this story where looking back at what happened I can see the perfect opportunities for me to have gotten out of this situation. I regret lying to my mom, I regret losing my virginity, and I regret—more than anything—that the incident at Point MacKenzie caused me to move toward Levi, instead of run away from him.

At the end of the conversation, Levi apologized and said, "We don't have to do it again until we're married."

Mysteriously, that's all I needed to hear. I was thankful he understood why I was so upset, and he seemed to totally respect my decision not to have sex again. His immediate agreement actually made me appreciate him more.

But I hadn't suddenly recovered from this traumatic event. Rather, I was reeling from a million emotions all fighting for attention in my mind. Later that day, my friend was at my house and we talked about all that had happened. It was a lot to process, so we needed to bring in more friends to help us talk it out.

"You need to come over," she said to another friend into the mouthpiece.

Do you remember when phones were "picked up" rather than "slid on" and were cradled in a receiver or hung from a wall instead of always living in your pocket or bag? Well, at my house we had a landline phone, and the kids were always spying on each other by listening from a different phone in other parts of the house. It was easy—you'd just carefully lift the receiver, put your hand over the bottom, and be very, very quiet.

Track had perfected this. I honestly think he could work for the

CIA or the FBI the way he always seemed to know my business. However, my friend wasn't aware of his tactics and didn't have a cell phone.

When she picked up the phone in the kitchen, she whispered to our other friend. She didn't realize that Track had suspected something was wrong in the house and had lifted the receiver to test his theory.

"Just believe me. You've got to get over here," she insisted. "Now!"

Apparently, my other friend at the other end of the phone knew something was up and wasn't going to come over until she knew exactly what had happened.

My friend finally relented. "Bristol and Levi had sex last night!"

Through the phone, my friend heard the loud *thunk* of Track slamming the phone onto its receiver.

"Bristol, I think I messed up," she sheepishly said as she hurried into my room. We looked out the window and saw Track furiously punching numbers into his phone and pacing back and forth in the driveway. Then came the yelling, "I'm gonna f—ing kick his ass," he told his friend.

Because I wasn't rowdy, Track was disgusted and shocked when he heard his hockey teammate had taken his sister's virginity. He never spoke to me about it directly; he drove straight over to Levi's house and threatened to . . . well, let's just say Track was an "abstinence only" advocate when it came to his sisters, and he was ready to enforce that philosophy with his fists.

Please don't tell Mom and Dad

I texted him. In retrospect, I underestimated my parents. When I finally did tell them that I was pregnant, they didn't come back at me with fire and brimstone. How much easier it would have been if I'd confessed to them before it escalated to pregnancy.

Only if you promise to never do that again!

he texted right back.

Though I'd later find out it was a hard promise to keep, it was an easy promise to make. I never ever wanted to have sex with Levi again until we were married.

I promise

I texted Track, and that was that. I didn't appreciate it at the time, but I'm now thankful to have a brother who cared about me so much.

A natural consequence of Track knowing about the Point MacKenzie incident is that Levi never showed his face around my house again. When we'd go out, he'd pull into the driveway, I'd hop into the car, and we'd go to his house or to a movie. (Girls, if your date won't go to your house and knock on the door, he's probably not the right guy for you.) Levi had gotten his license in May, which meant we had more flexibility and didn't have to rely on parents or friends to see each other.

And that's the story of how I began to live my double life and why Mom and Dad didn't worry about their straight-A, straitlaced daughter.

At least, not yet.

Chapter Four

Not Like Other Families

On November 7, 2006, we gathered at the Captain Cook Hotel in Anchorage with hundreds of volunteers, supporters, family, and friends who came in from the cold to hear the results of Mom's gubernatorial race. They'd been outside campaigning, holding the red signs on the side of the street, and praying that Mom would somehow pull it off. It had been a tough campaign for Mom, who was definitely the underdog, but as the precincts started reporting in gradually, we began to think, Could it be possible that she might win?

Mom said her thank-yous to all the supporters before we walked from the hotel to the Egan Civic and Convention Center, where the media had gathered to interview the candidates. I'm not sure how long the walk was—probably just a few blocks!—but it was the first time I'd ever worn heels, and it seemed like miles.

Plus, it was freezing cold and snow was piled up on the sides of the streets.

It didn't matter. We were so excited about the fact that Mom might become Alaska's youngest and first female governor, we could've run down the street. When we finally arrived, Leisl and my cousin Lauden marched into the center yelling, "Sarah! Sarah!" Mom went off to do radio and television interviews, while we probably acted rowdy in all of our giddiness. Finally, after Mom had given several interviews, the results of all the precincts came in.

I was standing with my cousins and family friends who had helped us with the campaign when I heard that my mother won the race; she ended up capturing more than 51 percent of the vote.

Everyone was so excited, and the atmosphere was electric.

It was clear right off the bat that Mom was not going to be "politics as usual." She was sworn into office in a hockey rink. It turns out the Carlson Center sports arena was the perfect place to accommodate the five thousand Alaskans who braved the cold and attended the event—including almost two thousand students who came on yellow school buses or traveled from nearby Denali Elementary. We got on a bus in Wasilla with every member of my family and made the seven-hour drive. Lauden and I put our headphones on and chatted the whole way. The drive was worth it. There was a lot of excitement all because of *my* mom.

She took the oath of office while placing her hand on the Bible Dad held for her. It was the Bible she'd used as a kid, and it had her maiden name—"Sarah Heath"—embossed on the front of it. Afterward, everyone clapped enthusiastically, and people started chanting her name from the cheap seats.

We were sitting onstage—five generations, from Great-Grandma Lena all the way down to Piper, who wore a bright red dress. What other color would do? Before we went out there, Piper wanted to wear a tiara on her head, so we let her attend our mother's ceremony wearing a crown! We were exhilarated as we heard the crowd roar, "Sarah! Sarah!"

Alaskan natives danced, bagpipe players played, and Jewel's dad sang. Iditarod champion Libby Riddles introduced Mom, since Libby was the first woman to win the 1,150-mile sled dog race. When she got up to the podium, everyone was still chanting, so she quieted the thousands by saying, "Okay, the governor said cease and desist."

It was a wonderful ceremony, which nicely honored my mother's accomplishment, showed she was definitely going to be a different kind of governor, and allowed us to celebrate with loved ones.

Levi wasn't there.

In fact, he never campaigned with us, never wrote out a name tag, and never went to any picnic or rally. I interpreted his lack of interest in the outcome of the race as a sign that he wasn't searching for fame. In fact, I actually found his lack of interest appealing, a weird kind of evidence that he was in our relationship for me and me alone.

When Mom became governor, it meant we had to make a lot of changes, including a move to Juneau. Though I thought I might miss Levi, I was ready for a change of scenery.

"Don't bring anything you don't absolutely need," Dad said to us girls as we excitedly stuffed our favorite clothes into our suitcases. It's rare to be able to start over in the middle of the school year at a new school, in a new city.

There was only one problem. Juneau is the most inaccessible capital in America. In the middle of the Tongass National Forest, you can only get there by air or sea. While thousands of people debark off cruise ships into the city, it's more complicated for people like my mom who are involved in the government. Some lawmakers have to travel a thousand miles just to get to the office, trips complicated by the lack of roads in or out of the city. (Juneau is the only capital not connected to the U.S. Highway system other than Honolulu, and the only capital to share a border with another country: Canada!)

It complicated my life as well.

"You're not taking your car," Dad said. He'd told me earlier that my first car would be my mother's black VW Jetta. It was diesel and a stick shift, which made it so cool. (We still drive it, and it's going strong at 180,000 miles and counting.)

"Dad, come on!" I protested. Mom was already preparing for her new role as governor, so this showdown was going down between Dad and me.

"If you're not playing ball, you don't need a car." As I mentioned, my enthusiasm for basketball had waned since middle school, and I was ready to hang up my uniform. But Mom and Dad believed athletics were one of the most important aspects of youth and weren't letting me quit so quickly.

"You're trying to bribe me?"

"You bet I am."

To get a car into the state capital, you have to drive it up to Canada and have it barged in. It was complicated, cost money, and Dad didn't want me to have extra time and a car in which to drive around and find trouble. I reluctantly agreed to their terms and stuffed my basketball shoes into my suitcase as well.

When we first arrived, we staked out our bedrooms. The Alaska Governor's Mansion is over fourteen thousand square feet, has eight fireplaces, and six bedrooms. The Greek Revival–style house is white, with large stately columns, black shutters, and a green roof. It seems like the kind of house that could be found in New England, except that there's a huge totem pole on the side. I bet our state might be the only capital with that decoration! Alaska, as you may know, is famous for these hand-hewn cedar poles that were used to communicate before natives had written language. The one at the mansion supposedly tells the story of how mosquitoes, sunlight, stars, tides, and marine animals were made. The stately home sits high on a hill, which gave us a really great view of the city. Plus, it was just a couple of blocks away from the Capitol building, the high school, and the downtown shops.

Though it was nice, it seemed more like a museum than a house. So we started moving in, Palin style. The most fun things we brought to the mansion were "buoy swings" (like tire swings but made from old buoys instead of Michelins), which we hung from a tree in the back. Mom also put a trampoline in the backyard. The transition to Juneau was pretty easy and was even more seamless because of the hard work of our house manager, Erika. It didn't hurt that she had fun boys our age. We loved hanging out with them. One of my fondest memories was when one of her sons and his friend jumped into the freezing river channel in the middle of winter wearing swim trunks. My friend and I were trying to drive back from the docks after we videotaped it. But after they stood on the side, held their breath, and plunged into the icy water, my car got stuck! So these two freezing cold boys in swim trunks and sandals had to push our car out of the snow before any of us could get out of there.

Dad still worked the North Slope, which meant he was thirteen hundred miles away. Track was traveling on a hockey team in Michigan, more than three thousand miles away. That meant that Willow, Piper, and I were the only Palin kids in the mansion. Talk about girl power!

One night, I was lying in bed around nine o'clock, when I heard a commotion outside.

"Bristol . . . Bristol!" I thought I was just hearing things until it lasted for fifteen minutes straight.

Reluctantly, I got up and looked out my window to see about a dozen high school kids chanting my name with a megaphone.

I opened the window and leaned out. "What?"

By this time, Mom had gotten out of bed, too.

"Hey, boys," she said. "Why are you yelling at my daughter?"

They explained that they were a high school group and were doing a scavenger hunt around town.

"One of the things on our list is to get a photo taken with one of the Palins," one of them said.

Mom loved it and thought I should not only let them take a photo but also join in the fun.

"You should join them, Bristol," she said. "Go do the scavenger hunt with them!"

That's when I knew Juneau was going to be an awesome place to live. I hurriedly got dressed and put on my shoes. However, after I got down to their cars, I realized that these kids didn't really mesh with me. Though they were incredibly nice, they told me they were on the "battle of the books" team and drove Subarus with "Going Green" bumper stickers.

Before we drove away from the mansion, another scavenger team arrived to get a Palin photo. Now, *that* was the team to be on.

The boys were rowdy, drove big trucks, and one of them even had a snowmachine in the back. In other words, they were my kind of guys. Because I'd told the first group I'd go with them, I did . . . but I was pressing my nose against the glass in regret that I'd committed too early.

Once we settled into a routine, our days went something like this. I'd go to school, then Mom would get the girls ready and drop them off at school. After school, Piper's school bus would drop her off next to the steps of the Capitol. She'd bring in her pictures, which she was forever making for people who worked there, who nicely hung them in their offices. Senator Menard even gave us a dog, whom Piper named Agia. Sounds like a nice name for a dog, right? Well, it was an acronym for Alaska Gasoline Inducement Act, Mom's signature project. But when we left, we gave that dog back! We also had a dog named Indy. (Well, we had three dogs named Indy. Indy 1 was a gift from Aunt Molly when I was two . . . right after the earrings. Indy 2 was a black miniature toy poodle that Willow neglected so much that Grandma and I gave it away when Willow was out of town. Indy 3 was the runt of a shih tzu litter, which didn't last long before we gave it to a friend, too!) Now I have a new dog named . . . Charlie. Yes, I did think about Indy 4 but decided to try something new. Maybe he'll last longer!

I see now that it was good for me to be back on the basketball team, even though it was a compromise. The basketball manager, Marissa, quickly became my best friend. Half Chinese and half native, she had dark hair and a funny sense of humor, and we became friends instantly. Since she was the manager, I asked her to hold my phone during our first game. Well, Levi just happened to call me during the game. And when I didn't answer, he called me again. And again. And again.

At the end of the game, Marissa handed me my phone and said, "Levi Johnston has called you forty times! What's wrong with this kid?" I remember she pronounced his name like "levy" and I laughed. "Is he crazy about you?"

"I think he's just crazy," I said, and we laughed.

We were instantly best friends, which made the irritations of basketball in an inaccessible city much more bearable.

In the lower forty-eight, mothers complain of their kids taking long road trips that get them home late at night and complicate homework. But in Alaska, sports require long road trips and even airplane rides to remote locations hundreds of miles away. Instead of quick trips to nearby towns, our "away" games took us to faraway towns. While we lived in Wasilla, most of our trips were manageable and within a few hours. But travel for Juneau basketball was a lot more difficult.

Competition required taking airline "milk runs," Alaskan Airline flights more like city buses than flights because they stop in every small town to let passengers off on the way to the destination. (The jets on the milk runs have large cargo areas sometimes loaded with Cordova salmon and other high-priced fish to take to Seattle.) We took about four basketball trips during the season, during which the whole team would fly into small towns like Sitka (in the southeastern part of the state that faces the Gulf of Alaska) for a game. Instead of staying at the high school or at a hotel, however, the entire team would be "housed out" with local families willing to let us sleep in an extra bedroom or, all too frequently, their floor next to the couch. Even when my mom was governor, I'd travel to remote villages, go to complete strangers' homes, and sleep on their floor. I think the only special treatment I got was once I was housed out at the home of the Ketchikan mayor, whose daughter played basketball. Most of the time, I was sleeping on whatever pull-out sofa or spare bed people had. Frequently, when people realized they were housing the governor's daughter, they'd stick a business card in my hand and ask, "Hey, we love your mom. Can you give this to her? I'd love to work for her on . . ." Otherwise, I was just part of the team.

We never housed out any of the opposing team's players at the Governor's Mansion, but it was completely open to our friends from school. When we lived in Wasilla, our place was where everyone always came over for food and television. My brother's hockey friends were always over, eating cookies my mom baked and annoying Willow, Piper, and me. People would always tell me how awesome and cool my mom was, and I'd readily agree. So I had the "cool mom," and the "intimidating dad." A good mix, in my opinion. In Juneau, even though the dynamics were different, with Dad working on the Slope, our open-door policy was still in effect. And it was a lot of fun to invite people over to a mansion.

Most of my classmates hadn't ever been inside of it, because governors never had young kids running around. There was a famous place in the mansion called the "cigar room," which became the hangout spot for all my friends. It had old leather couches, a wine cellar, a wet bar, and a card table. You just felt like you needed to smoke a cigar and have deep conversations in a cloud of smoke. That's not what we did, of course. Mom put an (more kid-appropriate!) air hockey table down there, and it became the most awesome place for friends. Also, because it had its own entrance, we could come and go without disrupting anyone else.

My bedroom was not as cool as the cigar room. The decor was stuck in 1972, with Pepto-Bismol pink walls, a floral print comforter with matching drapes, and other items we were warned we could *not* change. However, we did spiff it up a bit, with two new minifridges, a microwave, and pictures for the wall. Plus, it had three closets in it. One closet was full of jeans only while another was packed with color-coordinated, perfectly laid out T-shirts. It was a joy for a neat freak like me to have the space to organize everything in such a way. The bedroom had a balcony that stretched to Willow's room. During the summer, tourists were always coming by and taking photos, so my friends, Willow, and I would get out there and make noises at them, like bird calls, to see if they'd notice.

For the first few months, we had a chef; and the basement was like a Costco, full of soda, ready-to-eat food, and cookie dough, which made me very popular with friends at school.

I took advantage of it.

The first people I had over for a formal lunch date were Marissa and a guy named Hunter Wolfe.

I was introduced to Hunter the first week of school, and I loved his shy demeanor and dirty blond hair. He was into football, trucks, motorcycles, and dirt bikes. In other words, he was a typical Alaskan boy. But something that was not so typical was the fact that he treated me so well.

Once I left basketball practice and saw a note on my windshield.

Hey cute bball girl, hopefully we can hang out
soon!

"Check this out," I said to Marissa as I handed her the note. "No one's ever done anything this thoughtful before!" We were both surprised.

Sometime during the next few weeks, Hunter and I skipped class after lunch. I needed to go to the bank, and he came into the bank with me, even though we could've gotten into so much trouble for skipping school. Without even thinking about Levi back home, I immediately agreed to an invitation for lunch. It couldn't have gone better. When he reached for the check to pay, I noticed that he smelled so good. He was a gentleman, quiet, and very respectful.

After we went out to eat a few times, I invited him and Marissa to the mansion.

Blowing taxpayer money sure had its advantages!

"Hello," I said to our chef over the phone between second and third period. The school didn't have a cafeteria with long lines, milk in cardboard boxes, or hairnet-wearing lunch ladies. Instead, we were allowed to go off campus for lunch to eat where we wanted.

"I'd like to have some friends over for lunch, please."

"Wonderful," she said. Since Mom wasn't hanging out at the mansion all day, I think the chef was thankful to actually have something to do. "How many people have you invited and what should we serve?"

She could make anything for us—gourmet grilled cheese sandwiches, homemade tomato soup, Chinese pot stickers, Caesar salads, pizza, any variation of salmon, and . . . well, just about anything. The first day I had my friends over, I think we agreed on serving grilled cheese sandwiches and soup, and I anxiously awaited the opportunity to entertain them in our new glamorous location. When Marissa got out of Hunter's car, however, she somehow ripped her jeans all the way down the back. And that's how my elegant lunch began, with a big laugh, a rush to my room to grab an extra pair of jeans, and good food.

Governor Mom, however, considered the chef an unnecessary luxury. The cook was bored at the house all day, and Willow, Piper, and I never wanted any of her gourmet undercooked meat. Plus,

Mom thought it would teach us a bad lesson to always have a chef hanging around waiting to satisfy our every whim. And after she asked the government department heads to cut costs, she felt that we needed to cut costs, too.

After she unbudgeted the chef, we were sitting around eating moose hot dogs, and I couldn't hide my irritation any longer.

"Mom, seriously?" I asked, holding up the moose hot dog. Though those hot dogs are actually quite tasty, I complained that our family was getting fewer perks than previous governors' families. "We already don't fly first class, we don't have security detail at school, and we don't have people chauffeuring us around. Why can't we have any perks?"

Of course, she campaigned on fiscal responsibility, which meant making some drastic but important cuts. She replied, "Well, we're not like other governors' families, Bristol."

Only now do I appreciate her wisdom. It's hard to turn down things that make life easier, even if you know it's the right thing to do. Self-reliance is a virtue my parents tried to teach us in Wasilla, and they'd keep trying to instill it in us in Juneau. (They're still trying to instill it in us, come to think of it!) That's why I know that even if Mom one day becomes president, the Palin kids will still clean out their own trucks, shop at Target, and cook our own meals. Or in my case, defrost them.

When Mom had to host dignitaries for dinners, it required laying out a nicer spread than her famous moose chili. That meant that I had to help her when she didn't always have the help of a "First Spouse." I helped Mom decide on the catering menu, select floral arrangements, help the house manager, and select fonts for the place settings. But my normal life had very little to do with being the governor's daughter and everything to do with being a teenager.

And crushes are part of teenage life. One afternoon after

school, a bunch of kids were over and we were watching movies with friends. Hunter and I went to go get some food for snacks in the basement. When we got down there, he leaned over and kissed me! I had butterflies in my stomach because I knew that he was developing feelings for me.

This was confirmed when, not too many days after that, a floral deliveryman knocked on the door of the mansion with a bouquet of pink roses. I'd just been out sledding with my friends, so I was in the shower when the flowers came. One of my friends went down to answer the door. When she came back upstairs with the gorgeous bouquet, we all were amazed.

"They're for you!" she exclaimed.

I carefully opened the Hallmark card attached, which read, "When I wake up, I think about you. When I brush my teeth, I think about you. When I fall asleep, I think about you." And then, when I opened the card, he'd written a sweet note in his distinctive handwriting:

> *Ever since we started hanging out, I can't stop thinking about you . . . I hope you realize my feelings for you are true! Love, Hunter.*

I was absolutely blown away. It was the first time a boy had sent me flowers! I actually figured out quite quickly that getting flowers was kind of a hassle, because you had to take care of them until they inevitably died. Give me chocolate anytime over flowers. But nevertheless, I didn't know what to think of Hunter's kindness. Our relationship fizzled out over time, mainly because I didn't know how to respond to his kind gestures toward me. I'd always think, *Why is he treating me this way?*, or *What did I do to deserve this?*

But before that, I hung out with him, along with a tight knot of friends, including Marissa, Jacob, Susie, and Alex (Erika's son). Although we were all so different, with very different personalities, we had a blast. We'd go shooting at the range, which was a lot of fun until one day the police stopped us. We'd apparently been shooting too close to the road, which gave the cop a perfect reason to check my driver's license. I got a ticket because I only had a "provisional license," which didn't yet allow me to haul my friends without an adult. Oops! All of my friends' guns got confiscated, and their parents had to go down to the station to get them back. (I am no stranger to tickets, though. I've gotten tickets for speeding, for having illegally dark window tint, and other little things. In Alaska, like in many states, you get points for violations during the year. If you accumulate enough points, your license will be suspended for a certain amount of time and you lose your driving privileges. At one point, I had only four points out of twelve left!)

When my friends and I weren't shooting clay pigeons, we'd hang out and jump on the trampoline in the backyard, which seemed so out of place in the backyard of the Governor's Mansion. Once, it was raining and we were out there, jumping around and being silly. That's when Mom—the governor!—came outside, climbed onto the trampoline, and jumped with us.

We had so much fun in Juneau. After the cook left, we always brought up fun foods from the cigar room, and pretended to be chefs. Chocolate-covered strawberries was one of our favorite treats to make. We were always in that kitchen. Once, Willow was making everyone bacon, and the grease got too hot. When she burned the bacon, the smoke detectors went off and the fire department showed up at the mansion. It was a little embarrassing. (And this was the second time we had a visit from the fire trucks. The first time was when Mom tried to build a fire in the fireplace, only to find out the chimney was closed due to lack of use!)

We'd also entertain ourselves by repeatedly listening to rap songs, making fun of the lyrics, driving my car through car washes, walking around downtown Juneau, taking hikes in the mountains, and having bonfires where we'd roast marshmallows into the night. Sometimes, after Mom had gotten the fireplace working again, we roasted marshmallows right there in the mansion!

Some of the other people at school, however, didn't roll out the welcome mat. I'm not sure if they were intimidated by the fact that we were the daughters of the governor or they simply didn't like newcomers. They were mildly irritating at school, with the kind of petty viciousness only kids muster. They'd threaten me to stay away from their boyfriends, call Willow and me names, and gossip constantly about what we were—or weren't—doing.

However, one day things took a serious turn. Some of our new classmates posted an Internet threat against Willow. An eighth-grade girl told twelve-year-old Willow that her Samoan brothers were going to gang-rape her. It really unnerved Mom. Later, one boy posted something on MySpace about me: "Bristol's a slut when she's drunk and a slut when she's sober." My heart sank when I read that. I barely even knew who that guy was! That's how, at an early age, I began to develop tough skin and quickly get over all the untrue gossip about me and my family.

Though it was generally a wonderful semester, I was still pretty happy when summer arrived. My dad and I packed my bags and drove my car (by way of a ferry) home to Wasilla.

○ ○ ○

Alaskan summers are a welcome change from the doldrums of winter. During the school year, we'd go to school around nine o'clock in complete darkness, and by the time we get out of school around three o'clock, it would already be dark again. But during summers, it's pretty much light outside all of the time. And everyone takes advantage of the light. I remember how exhilarating it was as a kid to ask Mom, "Hey, can we go ride bikes?" Even though it was ten o'clock at night, she'd let us go by saying, "Sure, it's light outside."

Wasilla also never gets too hot, even though the new Wasilla Target has lots of swimsuits on the racks. In the winters, it's not uncommon for temperatures to reach twenty below zero. And that's not counting the windchill factor. (We rarely even talk about the windchill. That's for sissies. Basically, it's cold, it's going to be cold, and it will always be cold until the summer when it's slightly less cold.) That's why my hometown has so many little coffee shacks dotting the main roads and streets. It's not uncommon to see one, drive less than a mile, and see another. That's why I decided that learning how to make coffee might be a good way for a kid like me to make a buck in the summer of 2007.

Mom and Dad didn't force us to work. Though we weren't wealthy, my parents took care of our every need. Sometimes, just sometimes, they didn't see some of my wants as actual needs. For example, after I spent that time in Juneau, I was totally addicted to jeans. Seriously, I wanted to have every type in every color from

Nordstrom. But after Mom bought me a few, she was sick of shelling out the money for a thirst that was never quenched.

"If you want designer jeans, that's fine," she told me. "But you're gonna work for them."

That's why I got my first job that summer at an out-of-the-way café in Nordstrom's in Anchorage. It was a stylish, casual café with caramel-colored walls and cozy places to rest a tired shopper's feet. I got this job because by working there I qualified for the company's 20 percent discount on all of the clothes. What a better way to get more jeans? Of course, by the time I'd pay for the gas to Anchorage, pay for parking, and take advantage of that discount, I was not even breaking even.

Over the course of my teen years, my coffee experience helped me get other jobs at several of the coffee "shacks" around the area. The shacks are freestanding little drive-throughs that allow customers to pull up, get hot coffee, and—I hoped—leave me a tip. I served lattes, espressos, cappuccinos, and brewed coffee at the Sunrise Coffee Shack, which was owned by the wife of my dad's partner on the Iron Dog. Then I'd drive down the road to Café Croissant's little shack and pour coffee in the afternoons during the second shift there. Then, about a year later, I got a job working at the Espresso Café across the highway by my aunt's. Since it was all basically just the same job—smile, take order, pour coffee, take money—I don't think they were ever concerned about me sharing company secrets.

My first job, however, came to an abrupt conclusion when I got caught speeding and got a ticket on the drive to Anchorage. When Mom and Dad found out, they made me stop making that drive.

That was okay, anyway, because it allowed me to work at Nana's L&M Ace Hardware store in Dillingham, about four hundred air miles southwest of Anchorage. Every year, we'd make the journey down to Dillingham to see my dad's mom, Blanche, whom I've always admired. Nana has been in Dillingham so long, she's like a grandmother to everyone there and seems to be the glue that holds the town together. A long time ago, there was a perfume commercial that had a lady bragging that she could do it all, singing "Because I'm a woman, I can bring home the bacon, fry it up in a pan, and never let you forget you're a man."

I wonder what that commercial would be like if Nana had sung it. Not only can she hunt for her own food, catch thousands of pounds of salmon, and make clothing out of squirrel hides, she also sometimes breaks into prayer in Yup'ik (a language of indigenous Alaskans). She makes frying breakfast meat in a pan seem a lot less impressive, huh? Nana's store sits on Second Avenue, right in the middle of town, and when the commercial fishermen start showing up, her shop buzzes with activity.

The fishermen come every year, because the town is at the head of Nushagak Bay and the mouth of the Wood and Nushagak Rivers. These rivers have all five species of Pacific salmon—coho, chinook, sockeye, humpies, and chum—along with freshwater rainbow trout, Dolly Varden, arctic char, and northern pike. Plus, Bristol Bay has one of the largest salmon runs in the world. (This is ironic, since I'm named after this famous area, and I'm the only Alaskan who hates the taste of fish.) It's one of the most beautiful places in the world, as it sits at the edge of rolling tundra where caribou, moose, and bear roam through ridges of birch and spruce trees.

Everyone in Dad's family is in the fish business. My grandmother has a boat called *Bristol K.* My dad's cousin Ina lives in the village of Ekuk, where they run a fish camp and all of our extended family comes during the summer to fillet and smoke the fish. (In Alaska-speak, Aunt Molly is an Alaskan because she was born here. But my dad is a true Alaskan because his roots are so deeply here, and his family members are Alaska natives, belonging to the original people of our state. And one of the neat things about his heritage is that if you're related, you're family. There's no such thing as "extended family," everyone is simply a "cousin" no matter how far down the line.) Now that my brother Track is home from Iraq, he's taking over the family fishing business. It's a way to make money and to keep the family tradition alive.

The hardware store provides gear to Dillingham and even more remote native villages only reachable by water or air. The fishing season lasts about four weeks, so we'd stay in Dillingham and help Nana sell merchandise to the fishermen. Unfortunately, Willow and I weren't much help that summer of 2007. Did I mention we fight a lot?

One day we were cleaning the glass shelves under which guns, knives, and other valuables were showcased. She was on one side of the glass and I was on the other. She kept bugging me, so I took the Windex and sprayed it in her general direction. Even though I was a mile away from her, she immediately started screaming, "My eyes! My eyes!" Nana, sick of listening to our constant bickering, grabbed us both by the arm and said, "That's it! You're going home!"

As she shoved us into her van, the cold harsh reality sank in. My own grandmother had just fired me. (As I sit here and think about it, I doubt we were even getting paid. It takes a lot to get fired from a job you do for free, but we managed it.) This is the only job from which I've been fired.

The fishing season was still going, though, so Dad wasn't going to let us get away with being bad workers. The next night, we went out in his open-air fishing boat. There are two different ways to commercial fish in Bristol Bay. One way is to drift out in the ocean with nets behind the boat, and the other is to set nets on the beach. Because we have a permit that lets us use the required shore space, that's where we fish. So we set nets.

Okay, so I know this is a little harder than cleaning the gun cases at the hardware store or pouring coffee, but the way it worked was simple. We'd go to Dillingham, anchor our gillnet on the shoreline, and run it out a couple hundred feet into the water by a small boat called a skiff. When the salmon swam along the shore they were unknowingly right in the path of our net and didn't realize they were about to be caught and harvested for someone's dinner. When the fish put their heads into the mesh, the gills got caught in the webbing and they couldn't escape. (That's why the net is called a "gillnet.")

Alaska has some of the most stringent environmental laws on the planet, so there are only certain times you can go out to fish. The Alaska Department of Fish and Game looks at the tides and says, "You can only fish for three hours between the hours of two A.M. and five A.M." So that's what we did. We'd wait for the tide to go out and get on a boat in the middle of the night.

On that particular fishing night, it was freezing cold, and we had no pillows or blankets. (Dad is all business and never tried to make things unnecessarily comfortable for us!) Plus, there is no cell phone service at all out there on the water. At Nana's house, she has one home phone and one computer, but the only way to communicate with her is through a radio on the boat.

After we set the net, we had to wait. Because the boat was open air, there was nothing to stop the harsh wind from cutting right through us. There was a lot of wind and smashing waves, but it wasn't dark. It's rarely dark in the summers in Alaska, so we get to pack in a lot of fun (or, in my dad's case, work) at all kinds of hours. We were always pretty close to land, but I felt so isolated out there on the water. It was just me, my family, and nothing but waiting. It seemed we waited out there the whole night, and I gradually felt the tips of my ears, fingers, and toes go numb.

Eventually I had to take matters into my own hands. On the boat we carry "brailers," which are enormous bags in which you can put up to two thousand pounds of fish. Suffice it to say they don't smell too great. But I was desperate. I crawled right into one and got a little protection from the wind coming off the water.

That night, as I snuggled in that stinky brailer, I was wet and tired and couldn't keep myself from complaining.

"Come on, Dad," I begged. "Don't you care your oldest daughter is freezing to death?"

He didn't. We were out there to catch sockeye, we were out there to make money, and that's what we did.

Finally, we checked the buoys, and it was time to pull in the net. That's when I got out of the nasty, wet brailer and started picking the fish out of the net. Because the salmon swim in such tight schools, you don't have many "extra" fish that sneak in that you have to get rid of. To get the fish out of the net and into my former "sleeping bag," the brailer, I had to rip them out of the mesh and toss them into the container.

Not only is this our family tradition, it's usually a good way to make money, though it's a big gamble financially. Sometimes you make lots of money and some years you barely break even. A drift boat in a good season gets about a hundred thousand pounds of fish. These days, the price hovers around a dollar a pound. When Dad was Piper's age, the price was a dollar and twenty-five cents a pound. Even though that's a long time ago, it's how the price fluctuates. In other words, a good season depends on how many fish you catch and what the market is. If you get a hundred thousand pounds, but are only getting twenty cents a pound, there's no way around it. It's just not a good season.

"How many pounds did we get?" I asked Dad after we picked out the salmon.

He could tell I was busy calculating. "So it's going for sixty cents per pound, and I get 10 percent, that's . . ." Suddenly, I wished I hadn't skipped so many of Coach Brown's math classes.

"It's not much," Dad said as he skipped to the point. "But can't judge the season off one night, you gotta look at the whole season."

I remember being so glad to see the "tender boat" (the middle-men who buy fish directly from fishermen on the water). They tied up the boats, weighed and unloaded the fish, and took them to a shore-based cannery and processor.

But our part on the boat was done, and we went home to Nana's, where more work awaited. We washed the boat, packed the fish we were keeping for our family, and made sure we kept the truck and gear nets properly. In other words, summer "vacations" weren't really all that relaxing, but it did teach me the value of hard work.

But I still won't ever eat salmon.

Chapter Five

Failing the Test

'm having trouble admitting it here, but something else happened that summer of 2007 in Wasilla. Somewhere between all of the coffee pouring and fish catching, I broke my promise to myself and God.

I wasn't drunk, it wasn't an accident, and I did it on purpose.

After the original incident with Levi on Point MacKenzie, guilt settled around me like an unwanted friend, tagging along wherever I went, whispering to me in the night when my head hit the pillow. The way I dealt with it was simple. I tried to ignore the constant tugging at my spirit. And when that didn't work, I reasoned with it.

We'd messed up just once, I thought. I hadn't been sleeping around like a lot of my other friends. I wasn't known for being rowdy. Although we had an up-and-down type of relationship, Levi had committed himself to me, and we'd been dating on and off for one year by this time. Plus, we were going to get married anyway.

And that's what made things worse. I began thinking of a wedding in a totally twisted light. Instead of it being a romantic conclusion of a long dating experience or the poignant beginning of a long life together, it became my Band-Aid—the thing that would somehow make our bad decisions okay.

The side effect of grasping onto the idea of a future marriage is that I suddenly become very dedicated to the idea of a "happily ever after" scenario. So I ignored a lot. I overlooked his lies, his inability to attend school, and his occasional romps with other girls.

Yes, I knew about them. I saw them. But, again, I ignored a lot.

It wasn't like I had no other options. I was popular, had other guys asking me out, and—by this time—had already been on other dates and even had been interested in other guys. When I came back from Juneau at the end of the school year, I was driving to Aunt Molly's when Levi and I passed each other on the road. He was driving a new Silverado he'd gotten for his seventeenth birthday.

Hey, babe

he texted.

I just passed you. Pull over?

I pulled into the Yamaha parking lot, rolled down the window, and, when he drove up, I noticed he had more than just a new truck. There were hickeys all over his neck.

"What is all that?" I said, pointing to them. "Puke!"

"I was superdrunk the other night."

It doesn't sound like a good way to reignite the flame, but we did.

Believe it or not, we had a good summer. He was faithful to me during those two months, and things felt comfortable between us. We were inseparable. We didn't go out with friends, we didn't party, we didn't drink, and we didn't even really go out on dates. Even though we were so young, I felt like I was half of an old married couple.

Around this time, Levi started to spoil me materialistically. He gave me Coach purses, nice rings, Abercrombie clothes, as well as Coach and Juicy rain boots. It was nice to be treated well. He bought me whatever I wanted—and many things I didn't know I wanted—because he was trying to make up for straying so much in the past. Since he was trying to make things right, I had hope for our future.

That's why, so many months after Point MacKenzie, we had sex again. It was part "thank you," part "security deposit." After all, it seemed Levi had needs. If I wasn't going to fill them, I feared he'd go back to his old ways. And I hated the idea of him being with other girls.

After I started talking about this story on a national level— trying to get teens to think about waiting until marriage to have sex—every single reporter asked me, "Did you use contraception when you had sex with Levi?"

I think they desperately wanted me to say that no, we didn't. I think they wanted to find a chink in my mother's political armor, and to be able to say that my pregnancy was the result of my mother's old-fashioned values.

However, it couldn't have been further from the truth. Like most teens, society had taught us (wrongly) that "safe sex" would prevent pregnancy and heartache. So we used condoms.

I thought giving in to him would finally secure his attention on the way to the marriage altar. I thought he'd now think only of me. I thought it would get rid of the shame that followed me around day and night if I simply lowered my expectations.

And it worked for a while.

It was a pretty good semester. I only played indoor soccer league, which was different from my normal multisport level of activity. I worked at two espresso shops, and I was also elected copresident of the Prom Committee. That meant I was supposed to help select our theme, purchase the decorations, set up the photo shoots, and make sure that prom was one of the best nights of high school for our seniors. With my extra time, I also took online classes from BYU to get ahead in school. I wanted to get as many credits as possible so I could move forward in life. This was a slightly different approach than Levi was taking to high school. He didn't even attend, except for maybe a class or two at the alternative school to enable him to play hockey.

No one in my life supported my relationship with Levi. Track always looked like he was on the verge of kicking his ass, my cousins never understood what I saw in him, and my aunts didn't think he was worth dating. Mom and Dad, who had no idea we'd been dating so seriously for nearly two years, didn't like that I was spending so much time with him. Dad even offered one day to buy me a new truck if I'd just break it off.

I thought Levi was handsome, but my family would ask a million questions and make unflattering comments: "What do you see in this kid? He always has a fat lip because of all his chewing tobacco, he has a goofy haircut that looks suspiciously like a mullet, and he never goes to school!"

They were right, of course. I see that now.

As the end of the semester approached, Mom prepared to return to Juneau for the next legislative session in January.

"Don't make me go back there," I said. I loved my time in Juneau, but the thought of going back to that high school with the catty girls didn't sound too appealing. Plus, even though Levi had pledged to be faithful, I didn't want to be so far away from him.

"Well, you're not living alone here," she said. Dad's job on the Slope meant I'd be way too alone for a girl my age.

Thankfully, we always have family to rely on. When Mom asked Aunt Heather for help, she didn't hesitate to offer a place for me to stay—in her daughter Lauden's bedroom—about an hour up the road in Anchorage.

During the last week of school at Wasilla in 2007, I put one of my Coach purses down beside my desk in my human relations class and slid into the seat. My teacher, who was my eighth-grade basketball coach's wife, got up in front of the class and announced a new assignment. "Class, tomorrow I'd like for you to bring in five items that represent your life. They can be anything at all; I just want you to speak to the class about what those objects mean to your life."

I don't remember what the other four items were, but one of my items was a pair of scissors.

As I prepared to go to my third school in three years, I had another precious chance at a new start. I brought the scissors in because it represented a question I was struggling with in my mind.

I wrote, "Should I keep dating Levi, or should I totally cut ties with him?"

I may have made an A on the homework assignment, but I ultimately failed that test.

Anchorage is Alaska's largest city, and it was close enough to Wasilla that Levi and I could maintain some semblance of a relationship. Aunt Heather was like Carol Brady. She woke us up at 5:30 every morning with a hot breakfast and served a home-cooked meal at night . . . even though she worked a full-time job as a teacher of children with special needs. She and Uncle Kurt own the cleanest Chevron Gas Station in the state, so I enjoyed talking to him about business. In fact, I peppered him with so many questions, he let me borrow a book called *Rich Dad, Poor Dad*. Frequently, I'd bounce ideas off him about how to become financially independent through real estate and investing. They have three kids: Landon, their son who's the youngest in the family; Karcher, who has autism; and Lauden, who's one of my best friends. We always dreamed of being able to live closer to each other, and now we were sharing the same bathroom! It was like having a sister, which meant . . . twice as many clothes! We wore the same size, so we had fun trading out jeans and jackets.

I was nervous attending West Anchorage High School, which is one of the biggest and oldest high schools in Alaska. My first day there was even more intimidating than my first day at Wasilla High. As I walked through the doors, I saw all of the teeming people— natives, Samoans, African Americans, and Asians—about two thousand students packed into the school. Security guards stood at doorways, ready to stop trouble before it started. Everything seemed rougher, tougher, cooler, and much more exciting than my small-town life in Wasilla.

But Levi was never very far away.

Even though a short distance separated us during that semester, we made frequent trips between Anchorage and Wasilla.

And sometimes I went even farther to be with him. When his high school hockey team was playing in Homer, Mom wouldn't let me drive there with him. The trip would take two hundred miles, and she didn't want me to spend that much time with him. Instead, she suggested I fly and stay at a hotel with Ben's mom. But I was stubborn and urged Mom to let me go with Levi—I didn't want him to drive up there alone, I reasoned with her. Really, though, I simply wanted to be with my boyfriend.

I was so proud when she finally relented and said she trusted me.

We didn't leave Anchorage until eight or nine o'clock at night. It was March and the roads were icy and the snow was coming down. Instead of that worrying me, however, I was just so happy to be with Levi. Lauden and I always made fun of girls who sat in the middle of trucks to be next to their boyfriends. But on that trip I sat right there in the middle of his new red Silverado. He was so proud to have me in that truck, which he kept so clean. He'd put a sound system in it, painted the interior, and even put a sweet-sounding exhaust on it. He put his arm around me in that truck, and the world just felt right. I felt protected and loved.

Levi and I stopped for gas at a station near Girdwood, and we both got out to get pops from the store. At the same time, a few people were walking out, including a big guy who was strutting right toward us. Our fingers were intertwined. I remember that Levi was also holding a can of Copenhagen in his palm—and he just kept on walking. He stuck his chest out and walked straight to the door, right into that guy's path. When we got back to the truck, he said, "You can't ever let anyone get in your way. If you walk straight, they'll move."

I thought Levi was so tough, so wonderfully protective. I loved the feeling of being with him. We finally made it to Homer at one or two in the morning and went to the hotel Levi's team was staying in. Sammy, who was also there for the game, and I got a room there, too, and I put it on my debit card. I was proud to be able to afford my own hotel room because of the money I'd made at the coffee stand. After Levi got his gear out of his truck and we waited a long enough time for the coaches to be asleep, we snuck out to be with each other some more. That's when I knew I could marry Levi. Even though we'd just spent all evening together in the truck, we wanted to still see each other! I would've married him then if I could've.

The next day, Sammy and I drove Levi's truck to the ice rink. It was so cold inside the arena that the people in the crowd were all huddled up, trying to keep warm. I went out to Levi's truck between periods and got the only coat he had out there—a huge fleece pullover camouflage hunting jacket, which swallowed me up. When he went back to the locker room after the next period, he texted me saying that he loved the way I looked in his jacket. This brought a smile to my face, and I didn't care that I looked ridiculous in that coat. Levi was my man, and I didn't have to impress anyone else.

They won the game, which put Levi in a wonderful mood. Even the weather seemed to perk up! We drove home under clear, sunny skies. I sat right there in the middle of the truck, with Levi's arm around me, and everything was exactly as I wanted it to be.

Just a few weeks later, I was flush with the excitement of a relationship that was finally on track. I loved Levi, and loved watching him glide over the ice in his hockey games. And so, when I heard he had a game in Wasilla, I bundled up, jumped in my car, and drove the hour to the ice rink. I climbed up near the top of the stands and was happily chatting with my friends Sammy and Chelsea. They were filling me in on all the details of Wasilla High School life that I was missing. That's when my eyes landed on a girl a few rows down.

"Look," I said to my friends, pointing at the girl's back.

"What?" they said, not knowing why I was visibly upset.

"That girl is for sure wearing Levi's jacket!"

"How can you tell?" they asked.

"Because I'd know that jacket anywhere!" I said. "Trust me. It's his."

After the game, I got a ride home with Ben, one of Track's hockey friends. As we were pulling out of the parking lot, we drove past Levi and that girl walking out to Levi's truck. She was stepping on my territory. I was fuming.

Levi lied about it, but I knew what I saw. During that time, I started hanging out more with Ben. He'd stayed with my brother at our house while his parents were in the lower forty-eight the previous year. He was also a friend of Levi's. So when my heart was broken—once again—by Levi, he understood in a way that many people couldn't. After all, he knew all of the characters in my dramatic life pretty well. Though things didn't get romantic between us, I appreciated Ben's willingness to listen to my complaints and concerns. He became one of my best friends, a relationship that helped me make it through that semester of relationship ups and downs.

Levi was like an old pair of shoes I should've gotten rid of but kept around because of the comfort.

When, a few weeks later, a guy from West Anchorage asked me out I thought, *Why not?* He was a cute hockey stud, seemed very respectful, and I said yes. What did I have to lose?

He took me ice skating, and we laughed as we skated and I—

sometimes—fell on my butt, though he was there to make sure I didn't totally wipe out. While being out there on the ice with him, I had the same sensation I had when Hunter sent me roses in Juneau. Levi's idea of a fun date was watching television in his mom's basement. But this guy was polite and fun, and he showed me what dating's supposed to be like. I remember noticing how nicely he was treating me. Then, after the fun ice skating experience, he leaned over and kissed me.

It wasn't a big romantic deal, but I remember instantly being overcome with guilt about "betraying" Levi.

I never went out with that guy again. It was like there were invisible strings tying me to Levi, and nothing—not even those scissors I brought in to that human relations class—could cut through those bonds.

Though I missed seeing my mother while she was governing, Aunt Heather really stepped in and helped me navigate the difficult waters of high school—dating, peer pressure, and even the more embarrassing matters.

Then another thing happened that continued to bind me to Levi. Aunt Heather noticed that my cramps caused me such intense pain that I sometimes struggled to walk. It ran in my family, and every month I basically shut down.

"Bristol, we have to do something about that," she said. "Let's go see a gynecologist and find out if there's anything he can do to help with that pain."

Within a week, we were sitting in my doctor's office, listening to her describe the benefits of the birth control pill.

"Birth control pills help with cramps because they stop ovulation," she explained. "This decreases the amount of prostaglandins . . ."

I couldn't believe my ears. Many of my friends had to go to an awful no-questions-asked clinic for pills, but here I was with a legitimate way to get them. What a great development! This meant Levi and I could stop using condoms, and I could make sure I wouldn't get pregnant.

Mom called me from Juneau when Aunt Heather told her about taking me to the doctor. Though she was glad I might get relief from my terrible monthly pain, she had a not-so-veiled message for me.

"Now, Bristol," she said, "just because you're getting birth control pills doesn't mean you can go out and have sex!"

"Mom!" I said, totally embarrassed. "Puke!"

I felt like Levi and I had figured out a way to cheat the system. Sure, I knew it wasn't best to have sex before marriage, but I was doing the second-best thing. I planned on only having sex with one man my entire life. Since Levi and I were going to get married, I rationalized that our premarital sex wasn't *that* big of a deal.

Mom, however, was so oblivious to the hidden side of my life, she didn't see my total humiliation for what it was—a desperate fear she'd see through me. Instead, she figured her golden child was laughing along with her at the absurdity of such a thing.

My ruse wouldn't last much longer.

Chapter Six

Van Palin and Other Surprises

'm gonna go bear bait

Levi texted to me.

Want to come?

We loaded up our gear, got on a snowmachine, and headed out for the day.

Maybe you've never heard of this practice. In fact, you probably try your best *not* to attract bears when you're out in the wilderness. But up here, it's a common practice that starts long before hunting season. The way it works is simple. First you set up a huge fifty-gallon yellow drum (like the kind that holds fuel), then you strap it to a tree, and put a hole in the bottom of the drum. The hole has to be just the right size—not so big that the food falls out and not so small that the bear can't get his paws in. Then, you fill it with grease, cupcakes, stale doughnuts, dog food . . . basically anything around the house that you have to attract a hungry bear. Depending on what area you're in, there's a certain time to put the bait out before hunting season. Hopefully, the bears get used to coming by for snacks. By the time hunting season comes, you're ready for them.

Levi had already set all of this up by the time we arrived on our snowmachine. I checked out the big yellow drum, dumped some old food on top, and then walked the twenty or so feet to the tree stand.

Now, this isn't a rickety old stand, like some deer hunting stands. Rather, this is like a big tree house, with a screened-in window to watch for bears. Still, it was a little intimidating to climb up the handmade stairs on the tree—about fifteen feet in the air—and see the claw marks all over the stand.

After you fill your drum up with food and climb into the stand, all that's left to do is wait.

And wait.

Even if you aren't hunting, it's fun to be up there in the stand, right in the middle of nature. You can see moose, porcupine, and sometimes even wolves! So we carried our sleeping bags and a space heater up into the stand, leaned our rifles up against the wall, and whiled away the hours.

Levi and I were in a pretty good stage of our relationship. Even though I was in Anchorage, I'd see Levi every couple of days—any day I wasn't working late, I'd take a quick trip to Wasilla. Also, he seemed to be really into me (though I lived too far away to realize he wasn't actually being faithful), and I enjoyed seeing him during these trips. We drove home in the pitch dark, so the only light we had was from the snowmachine. I was holding on tight. As we drove home, Levi hit something—a snow-covered berm or something—and we wrecked. The snowmachine rolled over our legs.

As I got up, slowly dusting the snow off me and getting dirt and branches out of my clothing, I remember thinking, I wouldn't have wrecked if I was in control and was riding without Levi. After all, I had been snowmachining with my brother Track since I could walk. He taught me everything I needed to know about how to safely ride a snowmachine. I couldn't help but miss him at times like this. Track played on the elite Alaska All-Stars team before moving to Michigan to play hockey his senior year of high school. And it wasn't long ago that my protective big brother decided to leave Wasilla and protect our country overseas.

Track is the most private person you'll ever meet. Though he'd been talking about joining the military for a while, he hadn't mentioned it to me that past summer. I was blown away by the news then that he'd been talking to recruiters, the same ones who would come to the cafeteria at Wasilla High School to talk to potential recruits at lunch. I figured he'd stick around Wasilla. He was a full-time student at the University of Alaska at Anchorage, the largest college in our state, and he had a full-time job.

In hockey, he was the type of guy who was always getting injured, and finally it caused him to give up the sport he loved so dearly. He had even had shoulder surgery earlier in March of that year. I think that's when he started considering joining the army more seriously. He missed the camaraderie of the team and was sick of the way the war was playing out without him.

When he told Mom that he had been visiting recruiters, she was a little surprised. But Track had a very strong dose of the independent spirit my parents instilled in him, and he knew exactly what he wanted to do. He and his friend from high school, Johnny, enlisted into the U.S. Army as infantrymen when Dad was on the Slope.

Everyone was so supportive—yet emotional—about his decision. I knew he'd be a great soldier. He loves to help people, he loves his country, he's a very hard worker, he's very athletic, and he's really smart. What a perfect way to apply his skills by serving his country!

A few days later, after everyone had adjusted to the idea that he'd be gone, he went to his doctor to get his physical.

"A physical?" The doctor looked at him. "For the army? Are you nuts? You can't go to Iraq with your bad shoulder."

Track went from the high of deciding to join to the low of being told he wouldn't qualify.

But he is not easily dissuaded. He went straight to the military doctor, who looked over Track's forms, looked up at Track, and asked, "Are you good to go?"

"Yes, sir!"

He and Johnny took the oath at the military recruiting office in Anchorage on September 11, 2007, as my mom and Johnny's mom blinked back tears.

As his little sister, I beamed with pride at the thought of Track serving his country. But part of me wanted to just have him here in Wasilla forever. He had always been the kind of brother who watched out for his sisters and helped us in ways too numerous to mention.

The thought of him going to Iraq was especially unsettling, but it wasn't the only transition that lay ahead for my family.

○ ○ ○

When I was back home visiting Wasilla and having my bear-baiting outing with Levi, in early March 2008, Willow walked up to me, holding something in her hand. "Mom's pregnant," she said.

"Shut up," I said. "She is not."

That's when she told me she had found a black-and-white ultrasound picture. It was the first time I knew of my little brother.

"I found it on the table," Willow said. "They left some papers behind, and it was in an envelope."

"They were here all weekend," I said, trying to make sense of it all. "They didn't mention anything about a pregnancy!"

I immediately picked up my phone and dialed my dad, but he was already on the plane heading back to Juneau. My call went straight to voice mail.

"Dad, call me as soon as you get this," I said frantically into the phone. After I hung up, I turned to Willow. "I told you she was getting fat!"

In fact, I'd told my friends just a couple of weeks ago that I thought Mom was either gaining weight or was pregnant.

My guy friends protested; they always said she was hot, and they didn't believe she'd suddenly let herself go like that. And the fact that she had just turned forty-four made us wave off the idea that she could be pregnant.

But there I was with an ultrasound in my hand, trying not to be upset. As I waited for their plane to touch down in Juneau, I struggled with a wide array of feelings. Mainly, I just didn't understand the need for the secrecy.

Later I'd realize that Mom was worried that critics would complain that she wouldn't be able to perform all her tasks as governor with a baby on her hip. Only one other governor in American history has given birth while in office—Jane Swift gave birth to twins in Massachusetts. As the first female governor of Alaska, Mom worried voters would regret electing a woman into office.

Dad called me back as soon as he landed, and I realized there was more to the story.

"Yes, you are going to have a little brother, and he may have special needs."

His statement was very calm and very direct, but my heart felt like it had fallen out of my body. Not only was I dealing with the fact that Mom—*gross!*—was pregnant, I had no idea what to think about the fact that he might not be healthy. My cousin Karcher, with whom I lived in Anchorage, is autistic, but you can't tell by looking at him. Would my new brother look different than we do? Would he struggle for the rest of his life?

I started crying and clutched my phone so hard that it broke. My initial shock wore off quickly. By the next morning I was so excited to have a baby brother that I'd gone shopping for blue baby clothes!

The next day, Mom called three reporters with whom she had a good working relationship and asked them to meet her, Dad, and Piper at the Capitol. There was a big seafood reception for legislators put on by the city of Kodiak to celebrate its seafood industry. Mom thought it would be fun to break the news to these reporters before walking down to get some good lobster and king crabs.

But first, she decided to have fun with them.

"Hey, we're expanding," she said.

They were confused. Expanding government? That didn't sound like my Republican mother.

"No, the First Family is expanding . . . I'm pregnant. I'm having a baby in two months!"

The whole state was in shock—Mom didn't look seven months pregnant. However, the people of Alaska didn't respond as negatively as Mom feared they would.

KTVA11 did a news report on its six o'clock news, explaining how some people worried she wouldn't be able to fulfill her duties as governor. Reporters went to the street and asked people what they thought. The way they set it up, you'd think there would be angry voters shaking their fists in the air. So when they showed people from the street, it was almost comical. Every single person reporters interviewed said they believed it was great news. (One woman said, "If she can handle a state, she can handle a baby.")

It was only a few weeks later, however, that my aunt Heather got a call late at night. Mom had gone into labor early.

This didn't give me a lot of time to process all that was going on. On the drive to Mat-Su Regional Medical Center, I was almost overcome with worry about my new brother. Would the delivery be dangerous?

Dad, Willow, and I were in the room as Trig was born on April 18, 2008. When I saw his tiny body, I immediately knew that this little boy was perfect. As I held him, I noticed he didn't look like the rest of us. His neck seemed thicker than those of other babies I'd held. About thirty minutes later, Dr. Catherine Baldwin-Johnson came into the room. She was a longtime family friend, and we called her Dr. CBJ. She was the same doctor who delivered Piper years before and had made her a Noah's ark quilt.

When she came back into the room, Mom said, "Show the girls that line on his hand."

Dr. CBJ opened Trig's hand and said, "Now this deeper line is a characteristic of Down syndrome."

That was the first time we heard his diagnosis.

Down syndrome, I'd later learn, is a condition caused when a baby has an extra chromosome. This causes delays in mental and physical development. Although some kids need a lot of extra medical attention, others lead healthy lives. The miraculous thing about these babies—and children and adults—is that they see the world with delight and bless their families with unconditional affection. Even though Trig was just a few minutes old, I felt like he'd already changed me.

Dr. CBJ opened his tiny little palm and showed us a crease that ran horizontal on his hand. As we admired this new addition to the family, all of my cares and concerns evaporated. He was absolutely perfect, pure, and beautiful.

Which brings me to *his* name.

Believe it or not, Mom threatened to name him Trig Zamboni—after the ice resurfacers that clean and smooth the surface of a hockey rink. Thankfully (surprisingly), she chickened out. Like the other kids, his name is chock-full of personal meaning.

Trig is Norse for "strength" or "true," and Paxson is for one of Mom's favorite snowmachine areas in Alaska. Plus, my dad has an uncle named Trig. However, Mom did make good on a promise to make his middle name "Van." Get it? Van Palin, like Van Halen?

She even sent out birth announcements that looked like the cover of a Van Halen album, but instead of the iconic VH against the ring of fire, her girlfriend designed a VP. Seriously, you cannot make this stuff up!

The arrival of Trig confirmed to me that God knows what he's doing—he blessed us with a "perfect" child, allowing us to see firsthand that every baby is worth protecting and worthy of respect. Trig changed all of our lives for the better and solidified my pro-life stance. I also would like to share with others that you don't need to fear the news that you might have a special-needs child. It's a tragic fact that 85 to 90 percent of Down syndrome babies are aborted, and I have to assume it's because of fear of the unknown. If you could experience the joy we've found in Trig, that joy would erase any fear. In fact, we're proud that Trig is different and embrace and enjoy him so much more! We even have a bumper sticker that reads MY KID HAS MORE CHROMOSOMES THAN YOUR KIDS!

As I sat there looking at my new baby brother in the hospital, I'd never been more emotional or joyful in my life. Little did I know that in eight short months, I'd be lying in the same hospital room, in the same hospital bed, delivering my *own* baby boy.

I was more than a month pregnant, and I didn't know it.

My back was killing me. I was sitting in the coffee stand, and I thought I was going to die.

Maybe I'm pregnant, I thought to myself, though it was hardly a real possibility in my mind.

Then the next morning, Lauden and I went off campus for lunch, and I told her how weird I was feeling.

"I thought you were on the pill now," she said. "Isn't it taking care of your cramping?"

"I don't think it's working," I said, as I tried to calculate how long it had been since my last period and that day in the bear stand.

"I have ibuprofen," she offered, digging in her purse.

"No, I mean I'm afraid the pill *isn't working,*" I said again, and then more slowly, "I mean, that it *really* isn't working."

She looked at me sideways. "No way."

"Let's go buy some pregnancy tests."

We bought a pack of two and went back to Lauden's house.

The first one showed that I was pregnant. I thought it was probably a fluke.

The second one showed even more prominent pluses. But two tests aren't proof of anything.

I went to work after lunch, and on the way home, I bought a lot more tests . . . all different kinds.

We went back to Lauden's house, and I ran to the bathroom.

I went out to my cousin while we waited for the results. Two minutes never felt longer.

We looked at it. Another plus sign.

"Oh, my gosh, Bristol," she said. The news immediately sank into her.

It can't be right, I thought. *I've been careful. Maybe I didn't do the test right.*

With each new test, my heart got heavier and heavier.

And, by the eighth test, I was positively snapped out of my denial.

Two days later was Levi's eighteenth birthday, so I drove to Wasilla to see him. I was pretty nervous to tell him, because I didn't know what kind of response he would have. After a few minutes of trying to get my nerve up, I finally just blurted it out.

"I have something to tell you," I said. "Don't freak out."

Then, I showed him the test.

He looked up in disbelief. For a woman not in my circumstance that day, I can imagine that telling a husband he's about to be a father would be a rather emotional and amazing experience. Women think of all kinds of creative ways to reveal the secret. Some wrap up a baby bottle and give it as a present; others buy baby clothes and sneak them into their husband's bureau. Because this was definitely not a festive occasion, I didn't expect tears of joy. Which is good. Because this is what I got:

"Better be a f—king boy."

That's it. That's what he said. And the weird thing about it, I was glad his response was so favorable. He could've been angry, he could've demanded I have an abortion, he could've denied it was his baby. Instead, he seemed to see something good coming out of this less-than-ideal situation. He seemed to actually want a son.

I knew Levi wasn't the man he needed to be. He treated me terribly, he cheated on me all the time, he didn't work, and he didn't go to school. However, I felt this was the kick in the pants that would cause him to be different. Now he had no choice but to clean himself up and start providing for me and our baby.

Baby. Just the thought of it was so staggering. I wasn't afraid of taking care of one, because I'd taken care of Willow, Piper, and my cousins for years. The scary part was being a teen mom. Being alone. Being a disappointment to my family.

I took one look at Levi, on his eighteenth birthday, and I thought, *This baby and I are going to change this man.* I hoped one day I'd lead him to God, show him how to be a part of a family, and settle into some sort of family life together.

"I'll get a job," he said. "We can make this work."

So I nestled into his chest and willed myself to believe him.

O O O

My mother's baby shower for Trig should've been a time of celebration for me as we welcomed my new adorable little brother. Fifty of our closest friends gathered at our friend Barb's house to eat a potluck meal and a large sheet cake. It was white, decorated with blue icing in the shape of blue booties and a bib with a teddy bear's face on it.

Because Trig arrived so early, he was actually there, sleeping in my mother's arms, while she and my dad opened handmade quilts and tiny onesies. As they opened each present and passed them around for people to admire, everyone oohed and aahed. There's something wonderful about baby clothes and gear, how they conjure the idea of new life and starting out fresh in the world.

With every new present, however, I felt worse. *No one will throw me a baby shower,* I thought. *No one is going to be excited over my pregnancy.* So I tried to look happy—like a sister should—hoping the dread in my heart wouldn't come out on my face. I plastered on a fake smile, but every precious gift was like a stab in the gut.

No one could tell I was pregnant at the time. Interestingly, I was following in my mother's footsteps by hiding my pregnancy, but she had very different reasons. (While she was worried about maintaining her difficult job as the governor, I was worried about graduating from high school!) After all of her struggle with having a special-needs child, she finally came to terms with it all when Trig arrived. The baby shower was an opportunity to celebrate this new baby our friends only recently realized was on his way.

"Who in this room has the perfect child?" my mother asked as Trig slept.

Her question was rhetorical, a way of explaining that every kid has what the world might call imperfections. Truth be told, the Palin kids looked pretty together from the outside. Track was a well-respected hockey star, I was a good student, Willow was beautiful, and Piper was . . . well, everyone loved Piper. We all had a role to play, and I (the hardworking, no-nonsense oldest daughter) was about to add another descriptor to my reputation: unwed teen mom.

On that day, however, I wasn't worried about my reputation around Wasilla. I wasn't even really worried about how I'd afford diapers (those fears would come later). I was mostly worried about telling my family, who were gathered there in the home of a friend having a sweet time together.

Well, not all of us in Barb's house were. Willow, who was there with her friend, wanted a stick of gum.

Most people would've asked for gum, but my sisters are not "most people." They don't have what some psychologists describe as appropriate boundaries. They were forever grabbing my favorite pair of jeans, the best hairbrush, or nail polish or leaving the door to my room wide open. Willow and I were best friends 20 percent of the time, but the other 80 percent we fought like the worst enemies.

Always "in my business"—it was hard to sneeze without them knowing it—Willow stepped outside of Barb's house, went to my black VW, grabbed my bag, and plunged her hand into it.

But instead of pulling out a sweet minty treat, she found a pregnancy test. Not one, but eight . . . and all of them showed two pink plus signs. Note I didn't write "positive." Because I could see nothing positive about my newfound situation.

Willow's reaction was not a good indicator that things would go smoothly.

"What are you doing in my car?" I yelled as I came out of Barb's house with my friend Sammy. We were loading up the trunks of our cars with Mom's presents, and I was shocked when I saw Willow's head stuck in my car. "Get away from there," I yelled.

I didn't have to do an investigation to realize she'd found out my secret. The look on her face—of absolute shock—was enough to tell me all I needed to know.

We honestly didn't know how to break the news—we wanted to keep it that way at least until we figured out a game plan. We are a Bible-believing, Christian family. How would my parents react when they found out their seventeen-year-old daughter was sexually active? And, by the way, pregnant?

Willow flipped over the pregnancy test and saw the two pink plus marks. How did she react to the news that she was about to become an aunt? Did she congratulate me? Did her eyes fill with tears, knowing what a rough road I'd have ahead?

Hardly.

"I'm going to tell Mom!" she said as I grabbed the test and my brown purse Levi had given me for Christmas from her hands.

That's when I knew I could no longer keep my secret from my mom and dad.

"You might want to head over to the house," I said to Levi over the phone on the way home. "And you better be ready for this, because after today I probably won't have a place to live."

The drive, only about fifteen minutes, wasn't long enough for the amount of procrastination I wanted to work into it. I took the long way home, and on the way I called my cousin in Seattle who'd also gotten pregnant out of wedlock when she was eighteen. The daughter she'd had from that pregnancy is a beautiful little girl, the light of her mom's eye. I thought she might have some advice, so I dialed her number.

"Willow found a pregnancy test in my purse. And she's gonna tell my mom."

"I don't get it," she said.

"Kandice," I said with more than a whiff of impatience. "The test is *positive*."

"Oooohhhh," she said. Kandice knew the challenges that lay ahead. "Well, you better tell your parents today before Willow does. But make Levi do it."

"I'm going to make Sammy go with me so my parents won't yell at me," I explained.

"Smart," she said. "But I did it, so you can do it, too."

Her advice wasn't helping. My nerves were through the roof.

When we finally pulled into the long gravel driveway, my mom was happily taking all of the sweet presents from the trunk of her vehicle. My dad began mowing the lawn.

I sent Willow and Piper upstairs, took a deep breath, and—finally—began what was undoubtedly the worst moment of my life.

"Mom, get Dad," I said. "We need to talk."

We were sitting on a red leather couch. I was in the middle, between Levi and Sammy, and we looked like children who'd just gotten sent to the principal's office. Willow was upstairs, pressing her face against the railing, trying to eavesdrop.

"What do you want?" She laughed, assuming I was trying to pitch the idea of them buying a truck.

"Mom, this *is* serious," I said. "It's not a joke." When I spoke, my voice cracked a bit, and she knew something was wrong. "You're not going to be laughing and joking around after I tell you this," I said.

"Todd!" she yelled out the door. "Come in here!" He'd been mowing the yard, so it took him a bit before he came in the front door.

"What's going on?" he said.

By this time, I was sobbing.

Dad walked in and sat in the recliner. Mom sat on another couch.

"Well, you've got our attention," Mom said.

I tried to open my mouth to tell her, but it seemed that I physically couldn't. Tears ran down my face, and I felt like I couldn't breathe.

"Bristol, just tell me what's wrong," she said. "I can take it. Don't let us sit here guessing."

I had no idea what my parents would do when they heard the news, but I was less worried about their reaction than I was worried about disappointing them.

I tried to tell them once again, but the words got caught in my throat.

Sammy, seeing my parents' look of concern, put her hand on my knee as a signal that she was taking over the proceedings. Then, in one short sentence, my family found out.

"Bristol's pregnant."

"You're joking?" Mom said. It was a knee-jerk reaction to a totally dumbfounding statement.

"Would I be crying if I was joking?" I said through my tears.

They were completely and utterly shocked. I figured they'd never talk to me again. I figured I'd have to live in Grandma's basement or find a place with Levi. But instead of shame or blame, they immediately started asking me questions about my future.

"Okay," Mom said. "Let's figure this out."

By this time, Levi and I been together for two and a half years. "Well, I think we should get married," I said. Levi still hadn't spoken.

Mom nodded at this, but Dad—ever protective—wasn't about to let me settle.

"Whoa, whoa, whoa," he said, with his hand in the air. "Let's don't rush into things."

I was shocked that they didn't kick me out, and I was even more shocked that Dad didn't shoot Levi.

"Let's take things one at a time," he said. "What are you going to do about school?"

There in our living room, we talked about how on earth I'd be able to get my high school diploma while carrying, delivering, and taking care of a baby. We talked about Levi getting a job and how I'd deal with living arrangements.

In other words, we talked about my future. They were going to be a part of it, whatever it looked like.

And I was thankful.

Immediately, I started taking charge of the whole situation. First, I enrolled myself in homeschool for the first semester of my senior year through an online course. I also started thinking about learning a trade, the fastest skill that would allow me to take care of a baby. Even though this was less than ideal, I wasn't going to have my son bounced around without a stable family. I wanted him to have the kind of stability I had as a kid. I went to Walmart and bought a sweet crib, one that I had to assemble myself! As I put it together, I was overcome with anticipation about this baby. But I was also thinking that I'd just spent an entire week's worth of salary from the coffee shop to purchase it.

"I'm going to school," I said to Levi, who had already dropped out of school. I told him he needed his GED to get any sort of a job.

He agreed and went every day to the building where people prepare for—and take—the test.

One day he came home with a pleased look on his face. "I got it!"

I was so thankful that Levi came through with this one—rather big!—promise. Since Levi was a kid, he tagged along with his dad, an electrician, on wiring and construction jobs. I was thrilled when he got a job as an apprentice electrician in the Milne Point oil field with a major Slope contractor. Even though he worked three weeks on, one week off, I was so proud of him. The fact that he was gainfully employed made the time apart much more bearable. And he seemed to also be trying to cut back on his many expensive hunting trips.

During one day of the "three weeks on," we were talking on the phone when the subject of baby names came up. We'd always thought about what to name any future babies, but now we really had one to name.

"So what do you want to name the baby for real?" I asked.

Names, as I believe I mentioned, are very important to our family. "Okay. So if we have a boy, I'd love to name him . . . Blaine or Bentley," I said. I was a little nervous about sharing it with him, because I'd been thinking about that name already for years. "And if it's a girl, I'd like to name her Blaise or Keeley or Oakley."

After he suggested some names—all after popular gun brands—he said, "We're naming him Bentley."

From that point on, when he'd write me letters from the Slopes, he'd ask, How's my boy Bentley? He'd even sometimes write directly to Bentley, telling him that he'd already figured out what type of gun he'd buy him and what kind of hunting they'd do together. Once, he wrote, "Bentley, Mom and I will both try to give you a good life. Dad loves you more than anything. I'll be home soon to love on you! Take it easy on Mom."

       I was proud of Levi for having a good job and trying to perform well. When he came home after his "three weeks on," we had one of our typical evenings of watching television together. It suddenly became more memorable when he slipped a ring on my finger. Why do I describe it as "memorable" and not special? Because this is how it played out:

"This is pretty," I said, holding the ring up to look at it as I tried to process what was going on. "It's so beautiful."

"Yep," he said. "And it was expensive, too."

Not exactly the conversation a screenwriter would create for a romantic proposal, but it would have to do. Everything—and I mean everything—Levi did was half-assed. Whether it was school, hockey, jobs, relationships, or even proposals. After he gave me the ring, we simply went back to watching television. I didn't call anyone to tell them about it; I didn't text anyone. What would I say? There was no "story," or profession of love. There was nothing but a ring on my finger and a baby in my belly.

And the pressure of all these new developments was getting to Levi. For a guy who never attended high school regularly, it was a big adjustment for him to maintain a job, prepare for fatherhood, and get ready for a marriage, too. But I'd lost my patience with his slacker ways. So when he wanted to go sheep hunting, he tried to figure out a way to present the idea to me. It's not that I'm against hunting—far from it! But I'd lost my patience with how much attention and money he gave to these outdoor endeavors, and had hoped that he'd settle down a bit in preparation for fatherhood.

"I want to go sheep hunting. If I get a tattoo of your name on my hand, would you not get mad at me for going hunting?"

"You want to get a tattoo of my name?"

"Yes."

"On your hand?"

"Yep."

"All right, but I want to go with you to make sure it looks right."

Many people in Alaska have tattoos. Many get the outline of our state. Others get the Big Dipper and the North Star, as represented on our state flag. My brother Track has a tattoo of Alaska and the Statue of Liberty. But no one that I've ever met had a tattoo of the word *Bristol*.

That's how I ended up in a tattoo parlor, picking out the font for Levi's tattoo.

It was a long way from the Governor's Mansion, when the only font I had to pick out was for the seating arrangements.

Chapter Seven

Unconventional

To be honest, I wasn't actively following the presidential race of 2008. Though I knew John McCain was the GOP nominee and Barack Obama was the Democratic one, I was so obsessed with my pregnancy, my future, and—oh yes—Levi that I didn't care who should be McCain's running mate.

One day, a Wednesday in late August, Dad came home from the Slope early and announced Mom had left for a business trip.

Though Mom must've been excited about her journey, she didn't let on the night before she left. She simply said good-bye and walked out the door with her little bag like she'd done many times before. But this time, she was traveling not to Juneau or to Fairbanks. This time, she was heading to Sedona, Arizona, to John and Cindy McCain's place. Apparently, Senator McCain wanted to talk to her to determine if he should select her to run with him as the vice presidential candidate. They swept her away on a Learjet, with only our family friend Kris Perry in tow. Track, Willow, Piper, four-month-old Trig, and I had no idea where Mom was going, though a business trip was hardly going to raise suspicion in our house. The McCain staff had kept us in the dark so no one in Alaska would notice the Palin kids had all disappeared together when we left later to join Mom.

On the way home from picking Piper up from school, I remember a conversation we had in the car.

Dad was driving when his phone rang. He was talking to Mom, and when we got to the end of our long gravel driveway, he held the phone away from his mouth and said, "What would you think if your mom was chosen to run for vice president?"

"Yeah, right," I said. I probably was texting a friend or checking my e-mail. "That'd be cool though, Dad!"

I wasn't even paying attention to the conversation. It was so far-fetched that everyone brushed it off. It was the kind of casual comment I'd never think of again, except that the next forty-eight hours would be some of the most unusual of my life.

On Thursday, at about four o'clock in the morning, Dad walked into our rooms and woke us up.

"We're going to take a trip for something special. I want to take a trip to surprise Mom for our twentieth wedding anniversary."

"Where are we going?" we groaned, through bleary eyes. Our first guess was Hawaii, a much more common destination for Alaskans than many Americans since it's only a relatively short and inexpensive flight from here.

"It's a surprise," he said, though he looked like he was hiding something more than just the destination. "And by the way," he added. "I'm taking your phones."

Now, any parent knows that you simply cannot ask teens for their phones without getting some serious backlash. Plus, they never asked us for them unless we were on our phones too late at night, if we gave them attitude, or if we said a bad word. Not having a phone meant I had no way of telling Levi that I was leaving Alaska or even getting on a jet.

"Don't ask questions," my dad said. "If you don't want to come along without causing problems, I'll drop you off at Grandma's on the way. Who's coming with me?"

We started to pack, and Willow—reeling from the thought of having no phone for who-knows-how-long—sneaked and grabbed Dad's phone for any evidence of where we were going. Sadly, she couldn't find any e-mails or texts that gave clues about what was up.

"Pack nice clothes," he yelled from downstairs. "We're going to go out for a fancy dinner."

Of course, I didn't have many nice clothes that still fit. I was five months pregnant and hadn't told anyone but my immediate family, Lauden, and Sammy. I certainly didn't have cute maternity clothes that could replace my jeans and baggy sweatshirts. Though I loved my parents and wanted to honor their anniversary, this was a packing drama I didn't appreciate.

Willow and I packed Trig's diaper bag, then we loaded up, drove to Kris's house, and waited. I guess we had to go to her house first so no one would see Todd Palin's vehicle driving into a private airport in Anchorage.

"What's going on, Dad?" we asked.

He smiled and only said, "It's a surprise."

We looked at him blankly in exasperation.

"Dad, you're so phony. Tell us!" But he didn't blink.

"If you ask me again," he began, before issuing some sort of mild threat. I can't remember what he said at the time, but I do remember realizing we'd have to waterboard him to find out where we were going. (And knowing that Iron Dog champ, that probably wouldn't even work.)

Soon, a fifteen-passenger van arrived at Kris's house, already with a car seat for Trig installed. We drove to Anchorage in complete confusion, which only increased when we arrived at a private airport in Anchorage and got on a G5 airplane that was a lot bigger than the governor's small airplane. (Mom had put the governor's jet on eBay to save our taxpayers money!)

"Hello, Mr. Palin," the stewardess said. "Is this the first time you've ever ridden in a Gulf Stream 5?"

He acted like he did it all the time.

"This is the same private jet Britney Spears used a few weeks ago," she continued.

The force of taking off from the runway pushed us back into our seats. The pilot turned off the lights in the cabin, but there was no way we were going to sleep. There was a television mounted in the cabin so passengers could see where they were going . . . we watched it all day for clues to our ultimate destination.

"Have you taken out a second mortgage, Dad?" we asked. I knew he didn't play the lottery, so that was out. However, we couldn't figure out why the same parents who economized by clipping coupons for diapers and shopping for clothes at consignment shops would suddenly spring for a jet. It just didn't make sense.

Of course, we weren't the only ones in the dark. I later found out what all Americans now know. I was in the middle of what has been described as the best-kept political secret in the history of American politics. Once it became clear that John McCain and Barack Obama would be their parties' nominees, reporters scurried around trying to find out whom they'd pick for running mates. The names swirling around the Internet were the Democratic options like Hillary Clinton and Joe Biden and the Republican options like Rudy Giuliani, Tim Pawlenty, Mitt Romney, and Joe Lieberman. (Of course, Lieberman wasn't a Republican, but Senator McCain was a maverick, and no one ever knew what he was going to do!) Reporters slept outside their homes, watching for any indication that they might be in covert conversations with the nominees. News leaked that reporters were watching Joe Biden's mom's house, that the Secret Service had visited the home of Mitt Romney's sister, that the podium in Dayton where Senator McCain was going to unveil his choice was set at the height for a candidate who was approximately six feet tall, and that Tim Pawlenty had mysteriously cleared his schedule.

No reporters, however, were looking for clues in Wasilla.

I'd later realize that our departure that day would forever change our lives. As we sat in that awesome jet, our eyes glued to the GPS screen, we watched as one city would come and go. Another would come and go. Then, we were slightly shocked that we finally landed in . . . Ohio?

About an hour before midnight, we landed in Cincinnati and drove about forty minutes to a place called Middleton. We pulled into the parking lot of a redbrick hotel. The raggedy old hotel had dated furniture, small rooms, ugly pink walls, and an abundant supply of cockroaches.

I'd never even seen a cockroach before. Reporters might not think Wasilla is the prettiest town in the world, but at least we don't have roaches.

"Dad, this has gone on long enough," we said. "It's time to come clean."

We walked by some guys on the campaign who were dressed down and wearing construction hats. They had disguised themselves as a way of getting a large number of rooms without raising eyebrows. When we walked into a so-called suite, we saw Mom, wearing a pencil skirt and a blazer. Though we knew something very strange was going on, I wasn't prepared when a woman named Nicolle Wallace—dirty blond hair, dressed like a newscaster—stood before us and broke the news.

"Do you guys know why you're here?" she asked. "Senator McCain asked your mother to be the Republican candidate for vice president."

"What?" I said. I thought maybe I'd misunderstood. Even though Dad had playfully asked us about this at the end of the driveway, it had never entered our minds as a possibility.

"That's so cool!" Piper yelled. I couldn't believe it.

The room was full of excitement and so much joy.

Nicolle smiled and uttered one of the biggest understatements I'd ever hear, "Tomorrow your lives are going to change forever."

It was hard to sleep that night . . . and not only because I was worried the roaches might scurry over us. The next morning, the McCain staff decided they needed to get us out of the hotel before we were spotted. Of course, no one in Ohio would be able to pick the Palins out of a lineup. (And anyone in their right mind wouldn't assume that the next GOP vice presidential candidate would be staying in a hotel like that!) We got dressed from the limited options we'd hurriedly packed, and the campaign escorted us down the staircase in the back corner of the hotel. Then we piled into two white vans in a parking lot down the road. A black SUV with tinted windows was following us, and when we pulled into a parking lot behind a building, they came in behind us. Apparently, the Secret Service was about to meet the woman they'd have to protect.

Mom and Dad said their good-byes to us and jumped in the car with the agents, and we didn't see her until she was standing backstage.

The nation, and our family, was about to be shocked.

It was John McCain's seventy-second birthday, Mom and Dad's twentieth wedding anniversary, and a day after then-senator Barack Obama gave his charismatic speech accepting the Democratic Party's presidential nomination. We were preparing for my mom's big "reveal" but were still being hidden in the green room at the Ervin J. Nutter Center in Dayton, where fifteen thousand people were gathered to hear Senator McCain speak. All kinds of people were coming in and out, and the Secret Service was buzzing around. We were standing there with all kinds of people, including Senator McCain's daughter Meghan.

She ignored us during the entire visit, until an aide came in and said, "Meghan, did you meet Sarah Palin's kids?"

Only then did she brighten up, extend her hand to us, and smile. "It's such an honor to meet you. I'm a big fan of your mother's, and we're about to go on a fun journey together."

She seemed really nice, so I shook her hand and tucked away the sneaking suspicion that I might need to watch my back.

The green room was bustling with so much activity that I didn't have to spend much time with her anyway. Journalists kept coming up to us and introducing themselves and the papers they represented. It was a beehive of excitement and activity. Then, right before it was time to hit the stage, Mom whispered, "Say a prayer!"

Senator McCain walked out onstage, while we were still hidden from sight, to a podium that had a blue-and-gold Country First sign hanging on it. In the audience, there were a few signs that had only McCain on them. They were about to become outdated.

"My friends," he said. "I've spent the last few months look-ing—looking for a running mate who can best help me shake up Washington and make it start working again for the people that are counting on us. As I'm sure you know, I had many good people to choose from . . ."

I could hardly believe what I was hearing. Only a handful of people knew what this "maverick" was about to say . . . but I was one of the lucky few. I probably would've been more comfortable had I not been trying to hide my pregnant belly. Plus, it would've been nice to have at least a little more of a heads-up, so I could've gotten a nice outfit instead of the $10 dress I stuffed in the suitcase at four in the morning the day before. I realized I'd dressed Trig in Carhartt overalls and I regretted that he was making his national debut in them. I kicked myself for not packing better, but even though these weren't ideal circumstances, I listened and appreci-ated all that was happening.

"I found someone with an outstanding reputation for stand-ing up to special interests and entrenched bureaucracies," Sen-ator McCain said, "someone who has fought against corruption and the failed policies of the past; someone who stopped gov-ernment from wasting taxpayers' money on things they don't want or need. . . ."

I noticed he didn't use the word *she* or *woman*. To do so would be to instantly eliminate most if not all of the other candidates who were talked up as options.

I thought about my grandpa and grandma, and how I wished they could be there with us. I wondered if they had any idea that they were being talked about in a speech in Ohio by the Republican nominee for president! (I'd later find out that the family back home had been alerted around 5:30 in the morning, Alaska time, to turn on their television sets. They were completely shocked!)

" . . . They taught their children to care about others, to work hard and to stand up with courage for the things you believe in."

It was a surreal experience to be talked about from the lips of Senator McCain. Though Mom was governor of Alaska, I never felt like the eyes of the state were on us. We were just locals like everybody else. But this? This was hard to process. Especially when you have a little morning sickness. Plus, I was worried I wouldn't be able to respond to Levi's texts!

"The person I'm about to introduce to you understands the problems, the hopes, and the values of working people; knows what it's like to worry about mortgage payments and health care, the cost of gasoline and groceries. And, I am especially proud to say in the week we celebrate the anniversary of women's suffrage, a devoted wife and a mother of five."

When he said "women's suffrage" and "mother of five," the crowd interrupted him and started cheering. American flags waved.

"My friends and fellow Americans, I am very pleased and very privileged to introduce to you the next vice president of the United States—Governor Sarah Palin of the great state of Alaska."

I can't tell you how much pride I felt at that moment. At several points in my life, I'd look at Mom and just be amazed at her composure and coolness under all kinds of pressure. As I listened to Senator McCain describe her, obviously, I experienced one of those moments. But I didn't have too much time to process the coolness of my mom. Suddenly, this majestic music started blaring over the loudspeaker (it sounded like the sound track of a boxing movie, when the characters are just about to have an exciting fight). We took a deep breath and started heading to the stage. It was like going out onto the field during the Super Bowl. The yells and the chants from the crowd were deafening. We walked under the bleachers, into a special entrance, and finally emerged onstage. I was holding Trig in a white blanket, which nicely hid my stomach. Thankfully, he was totally passed out.

We all waved to the crowd, smiling and trying to take in this very strange experience. Mom took some time to genuinely soak up the crowd's excitement. After the family sat down in the chairs behind the podium, my mom thanked the audience, Senator McCain, and Mrs. McCain. "I will be honored to serve next to the next president of the United States."

Everyone erupted! But, of course, it was not to be. At the time, we had no idea how walking out on that stage would change our lives forever.

It was probably a good thing we didn't know.

Chapter Eight

Looking the Part

Though I was thrilled our lives had suddenly been put on such an unexpected path, I didn't yet feel completely comfortable being so . . . public. After all, I'd been hanging out at my house for several weeks, not wanting to deal with the idea that people might detect my pregnancy.

In one regrettable moment when I did venture out into public, I held Trig in my arms at one of Willow's basketball games. An old family friend came up, looked at Trig, and quipped, "Isn't having a baby around a good form of birth control?"

Um . . .

Of course she had no reason to suspect I was pregnant. Heck, some of my extended family didn't even know. And America certainly hadn't found out.

At least, not yet.

○ ○ ○

As soon as Mom was revealed to be Senator McCain's running mate, we were sucked into a whirlwind of frenetic campaign activity, a fun temporary distraction from my pregnancy. After that Dayton speech, we got on the campaign bus, a really fancy rig.

As we climbed on with our diaper bags and baby paraphernalia, Dad whispered in his sternest voice, "Don't touch anything!"

That's when Cindy McCain saw us coming on and piped up, "Bring me that baby! Piper, help yourself to cookies or whatever food you can find."

The bus was sandwiched into a long caravan of about thirty-five cars, SUVs, and vans and began a whirlwind series of campaign stops.

Amid all of the handshaking and the baby kissing, Mom realized that she was just about out of diapers for her own baby.

You know how it is on a road trip, when you want to stop for a bathroom break and the driver grumbles? Imagine what it's like to stop a caravan of more than thirty vehicles? And imagine that you are suddenly the most talked-about, controversial person in America and you just need to pick up a few things in aisle 9 of Walmart.

The first thing we realized was that our motorcade had grown from about thirty to perhaps more than a hundred cars who were following us down the interstate to get a glimpse of the action. When we pulled over, *everyone* pulled over.

Mom, Trig, Aunt Molly, and I went into the store, with Secret Service agents following us down every aisle. As soon as we walked in, it was bedlam. People rushed up to her and she ended up getting swamped by well-wishers.

I don't think anyone anticipated how much excitement and energy my mother would bring to this ticket. All throughout the campaign, people not normally interested in politics would show up from all around and wait for hours just to hear her speak. Willow and Piper would go out onstage with her during campaign speeches and frequently go out into the crowd and "work the rope line" afterward. That simply means they'd smile, sign autographs, shake the hands of supporters who wanted to catch a glimpse of my mom, and pose for lots of photos. I didn't do any of that. After all, I typically was wearing a comfy sweatshirt to hide my stomach. (Though to be honest, this is what I prefer to wear even now that I'm not pregnant!)

It put me at odds with others on the campaign trail.

"Meghan," an aide had said to her when we were waiting together in a green room in Dayton, "you look beautiful."

She rolled her eyes, put her hand on her hip, and said, "I should. This dress cost a thousand dollars."

I didn't volunteer that my dress cost less than two value meals at McDonald's, but I did realize I'd never met people quite like this. After all, Alaskans are more concerned with how warm clothing keeps you than the label in the collar. Perhaps I was being too sensitive (those pregnancy hormones, you know), but I felt people looked down on me because I didn't "look the part." My pregnancy made me look fifteen to twenty pounds "overweight," I had limited wardrobe options, and I didn't really want to be in the spotlight.

When we finally wrapped up the first leg of the campaign trail in Ohio, we hurried to the location where the GOP convention was to be held in just a few days.

But first, I had to do something.

O O O

"Grandma," I said into the phone. "We have to talk."

Though I wasn't sure about the best time to reveal my pregnancy, I knew I only had a little time before it became too obvious to hide. However, there were a few people I admired so much that I could barely stand the idea of telling them. First, I didn't want to disappoint my grandmother, the former Catholic turned evangelical who was always very serious about her faith . . . she was somewhat of our family's spiritual matriarch. Second, I didn't want to tell my aunt Heather, who's so much like a mom to me that I lived with her family in Anchorage while Mom served in our state capital. And last, I didn't want to disappoint another "momma figure," my mom's best friend, Juanita.

As I held the phone up to my ear and heard my grandma's voice, I knew it had to be done. However, I couldn't force my mouth to form the words. My mom's friend Kris, who was with us on the campaign trail, was sitting beside me and saw me break into tears. She grabbed the phone from me and told Grandma the simple, yet devastating to me, news.

"Bristol's pregnant."

To her credit, Grandma took it well. I remember her voice shook as she said, "Now, now, now . . ." (Do all grandmas say this while shaking their head, or just mine?) "This is going to be harder than you think."

She'd raised four kids herself and could see the difficult path that was ahead of me more than I could. But even though I could tell that the news was pretty jolting for her, I felt relieved that—at least partially—my secret was no longer mine alone to bear.

I didn't know my secret was about to be shared with an international audience less compassionate toward me than Grandma.

The Republican National Convention was held at the Xcel Energy Center in Saint Paul, Minnesota. We stayed at the downtown Hilton in my mother's gigantic suite that had two bedrooms. Mom, Dad, and Trig stayed in one bedroom, and we sisters stayed in another. (Track and some of our cousins stayed down the hall.) I'm happy to report that—now that we weren't in hiding—there was not a cockroach in sight!

It seems I wasn't being too sensitive, after all, when I felt others considered my wardrobe unacceptable. Because there in the middle of the "living area" of the hotel room stood racks and racks of clothing. Skirts, tops, dresses, men's clothes for Dad, and a whole section for us girls. In the corner of the hotel room, a sewing machine sat with a seamstress who was ready to tailor the clothes to fit us perfectly. We didn't ask for any of this!

When we walked in and saw all of the clothes, Willow and I just laughed. The McCain campaign had flown in a bunch of stylists from New York—people who'd never met us and knew nothing about us—to help us look our best. Or, at least, to "look the part."

I am a conservative dresser, so as I looked through the rack I saw item after item I'd never normally be caught dead in. There were gowns, diamond earrings, pearl necklaces, Gucci shoes. I know it sounds like a girl's best dream, but we were accustomed to fleece pullovers and jeans. It's not like I didn't get the enormity of the situation. History was being made: for the first time in American history, the GOP had nominated a woman to the national ticket. Appropriate clothing was necessary and appreciated. However, shoes this expensive were a little bit of a stretch, even for history.

On the rack, we even found a cashmere sweater for Levi.

Let me put those two words together again, to allow them to sink in: Levi. Cashmere.

The stylists, who apparently worked for celebrity newscasters, helped us look through the clothes to find something nice to wear for Mom's big speech at the convention.

At any rate, I was thankful to have my entire family in Minnesota. Aunts, uncles, cousins, and in-laws were so supportive of my mom that they loaded into airplanes and made the trip to Minnesota just to show their love. Even Levi came, though he was really sick as my aunt drove him to the plane. He had to actually ask her to pull over to the side of the road more than once. Plus, they had to fight bad weather to arrive safely. On September 1, Hurricane Gustav was about to hit the Louisiana coast, causing all kinds of travel complications. (That's why Senator McCain canceled the first day's activities, other than a small opening ceremony.) By the time everyone arrived at the convention, they were simply overjoyed at being a part of it . . . and Levi had calmed down enough to enjoy the festivities and thank us for letting him join in.

The joy would not last long.

Mom was brushing her teeth in the hotel suite when she saw something quite disturbing scrolling on the television. It was an unapproved message from her that read "Gov. Sarah Palin on teen daughter's pregnancy: 'We're proud of Bristol's decision to have her baby and even prouder to become grandparents.'" She was shocked to read it.

We were all shocked. Good thing I told Grandma the night before!

For some reason, I didn't realize the ripples my pregnancy would cause throughout the nation. I was thinking about what kind of schooling I needed to make the most money in the shortest amount of time. I wanted to learn a trade. I was in "survival mode" and didn't really consider how my own personal drama was playing out on the national stage or think anyone would necessarily care about it. The one thing that was embarrassing is that I couldn't hide the fact that I'd royally messed up. Many sins can exist under the surface, undetected by others, allowed to grow and fester without anyone knowing. Pregnancy, as difficult as it is, has a way of forcing you to deal with your sexual mistakes out in the open, demanding a frank honesty that other sins don't. (For example, the effects of sins like gluttony, greed, jealousy, or hate take longer to show up—if ever—so people are less likely to come face-to-face with their flaws.) But there I was, after months of hiding, outed as a seventeen-year-old unmarried, pregnant girl.

Think you've had a bad day?

Immediately, the news of my mistake circulated all around the country and the world. It would've been one thing had they reported it accurately. Even the hard truth is still digestible compared to the unbelievable lies that began getting airtime. Some people actually suggested that Trig wasn't our brother. They even insulted us by suggesting he was Willow's baby. Or mine! It was so stupid, I couldn't believe we had to respond to it. But there we were, at the GOP convention, fighting off some of the weirdest and most malicious accusations.

I look back on that and wonder how we made it through that time so well. But I don't think about that for long. All we had to do when things got tough was talk to Aunt Molly and Aunt Heather, who'd traveled so far to support us. Or I could hug my grandpa's neck, or exchange sarcastic glances with Willow. Family means everything to us, and no matter how much the media tried to tear us down, their powers of destruction couldn't compete with our family's ability to cheer us up. Even having Levi there was a huge source of comfort. I wasn't blind enough to still think he'd help provide for me and the baby, but the trip to the convention was so new and fun that it helped temporarily blind me to our underlying—but very real—problems.

The side effect of everyone on the planet knowing I was pregnant was that I could finally reveal my cute "baby bump." As we prepared for Mom's speech, I chose a simple gray dress with cap sleeves, which was the first thing I'd worn that didn't try to hide the little bundle I was carrying. After I got my ensemble on, I checked myself out in the mirror. Other than the fact that I'd never buy a dress so expensive in my life—and not that one!—I had to admit, I felt happy that I no longer was always keeping a secret.

For some reason, the presidential candidate is supposed to arrive at the political convention only after everything has started. I guess it's kind of like going to a party thirty minutes late. It's just what you do.

When Senator McCain arrived in Minneapolis from Cleveland, we met him on the tarmac. After the stylists got us dressed and made sure we looked presentable, we arrived and realized that this was a very big deal. Photographers were everywhere, and news trucks (with their satellites on top) surrounded us to capture the event. Reporters knew a good story when they saw it: War Hero Meets Boy Who Knocked Up the Vice Presidential Candidate's Daughter. It was too much for hungry reporters to refuse, and that's exactly how they reported it.

Right as we were going out onto the tarmac at the airport, we met up with the McCain family. Meghan, who was wearing a cute gray dress and leggings, took one look at us and said, "Well, if I would've known we were supposed to dress up, I would've!"

Every time we saw Meghan, she seemed to constantly be checking us out, comparing my family to hers, and complaining. Oh, the complaining.

We stood for just a few moments before we were told to go out to the tarmac. Cindy McCain was walking with me and pulled me aside to talk privately. You can probably tell from television that she looks like a queen and holds herself like royalty. Her bags were so expensive. I've never seen people with so much Louis Vuitton luggage, so many cell phones, and so many constant helpers to do hair and makeup. When I saw their wealth, it was hard to imagine we'd have anything in common to talk about.

"Bristol, I have three things I want to tell you," she said. "I just want you to know that I want to be one of the first people to hold your baby. Also, I want to go to your wedding when it comes together, and lastly . . ."

She paused just a second, and added, "John and I want to be godparents of your child."

I can't even remember how I responded. I think I may have laughed it off, unsure if she was serious. I had just met her, and I wondered why she wanted any type of guardianship over my child. I was nice to her, but I was left speechless by her comment.

As we continued onto the tarmac, I looked up and saw snipers on the roof. There were also serious-looking Secret Service men and police dogs everywhere. While Mom had minimal protection while she was governor, the men protecting John McCain—and now us!—were many, tough, alert, and intimidating.

When Senator McCain got off the plane, he hugged Mrs. McCain, Meghan, his daughter Bridget, and his son Jack before turning his attention to us. He graciously greeted my mom and my dad, and then he turned to Levi and me. I stood to my mom's right, and Levi was to my right. Senator McCain greeted us warmly and shook Levi's hand. Then he said, "Let me see your hands again." Levi showed him his hands, and Senator McCain said, "Those are workingman's hands."

Senator McCain, I thought, *you've got that one wrong.* (That's okay. I thought Levi's outdoorsy exterior and love of hunting, fishing, and chewing would translate into actually keeping a job too.) Nevertheless, being there with him by my side was a remarkable experience. It was our first trip together, and we had so much fun doing simple stuff like ordering room service with our cousins. His presence, though obviously imperfect, was better than being alone and wondering which of my friends he might be cheating on me with back home. Standing there on the tarmac, with possibly the next president of the United States, somehow legitimized us as a couple and painted a picture I'd hoped would become a reality. And so I stood there on the windy tarmac and was thankful.

Though it was immortalized by all of the photographers, it was over very quickly. On the way off the tarmac, my mom held Levi's arm and said, "Levi, doesn't Bristol look beautiful?"

He looked at me from head to toe, and simply said, "Yeah."

I know it doesn't sound like much—it wasn't—but it was the first time he'd told me I looked pretty.

Then we rushed to get back to the hotel so Mom could put the finishing touches on her speech. A great deal was riding on it, after all.

On Friday when we stood behind her on the platform in Dayton, we had no idea the creative lies the media would brew up over the weekend. The slur that she wasn't Trig's real mom was just the tip of the Alaskan iceberg. Left-wing bloggers and even "reputable" news agencies started shaping a story about her and my family that couldn't have been further from the truth. She was called a religious radical who wanted to mandate creationism and abstinence-only education be taught in public schools, and they lied that she slashed funds for a program benefiting pregnant teens, banned library books, was a Pat Buchanan–supporting Nazi sympathizer, had an affair, arrested Alaskan women who'd been raped and had abortions, and—for good measure—encouraged Alaska to secede from the rest of America.

Also, remember Uncle Mike? Suddenly, stories about that stupid Taser incident started bubbling up again. Not as an example of how our family tried to help Aunt Molly. Instead, the news organizations were wrongly accusing my parents of trying to get him fired from his job as a state trooper. It was an "abuse of power" story, one of the worst kinds when you're running for office. The press even gave it a -gate suffix, which no politician ever wants. They called it "Troopergate," though in Alaska we called it "Tasergate."

Because none of these accusations were actually torn down by the McCain campaign, people began thinking that she wasn't qualified to be vice president, that she was McCain's "Hail Mary pass," and that she would wilt under the heat of the national spotlight. By the time the GOP convention started, Mom's media image had been so twisted that we looked forward to her actually being able to speak for herself, which she was used to doing and excelled at.

As Mom prepared for the speech, we were new, unknown, from an exotic locale. And wow, what a speech . . . Even though it was natural for us to be getting media attention, it seemed to be hard for Meghan. This tension finally came to a head.

On the evening of September 3, we were preparing for Mom's speech. Every time we went out into the convention center, we were to look camera-ready. The McCain campaign believed that packaging was one of the most important aspects of our big debut at the convention, and all of us dutifully went along. Two makeup experts and hairdressers were available to us on the floor of our hotel, and they were in charge of spiffing us up. On the night of Mom's speech, I was sitting in the chair getting my hair done with Willow, when a wet-headed Meghan McCain stormed into the room. She was late and needed someone to help her.

She looked at all of us sitting there getting our makeovers, GOP style. (Hey, that sounds like a good idea for a reality television show! What about it, TLC, now that Mom's Alaska show is over?) But as Meghan looked at us, you could tell she was supremely irritated.

"I need to be worked in," she barked.

One of the stylists took a bobby pin out of her mouth and apologized. "I'm sorry, they are all waiting." She motioned to Willow and me, and then to Piper. There were only two hairdressers and a lot of hair.

All of a sudden, Meghan's face registered a wide array of emotions. In a five-second time period I saw her face go from resentment to anger to bitterness back to resentment and then—finally—to rage. However, in a display of temporary restraint, she pressed her need more urgently.

"I just need a blowout and some makeup," she said once again. "I'm running late."

"I'm sorry," the stylist replied, saying her words carefully as she sensed the pressure rising in the room. "It's just that they'll be getting more airtime than you will."

In the chaos of a campaign trail, you rarely find silence. But there, in the hair and makeup room, we found it. For about five seconds, everyone sat slack jawed, waiting for Meghan's response as she pondered the fact that we would be on camera more than she would. We were not disappointed.

"If anyone had told me that I had to do my own hair and makeup," she screamed, "I would've done my own f—ing hair and makeup!"

You could tell as she yelled there was something quite complex going on inside her. Or at least I liked to think so. After all, nerves were frayed and the stakes were high. Surely she wasn't so self-obsessed that she believed everyone else should scoot over so she could take priority?

Willow didn't know if she should giggle or gasp at Meghan's reaction. Britta, who was Track's longtime girlfriend (now his fiancée), had come in to borrow some hair spray and witnessed the whole thing. After Meghan stomped out of the room, Britta pulled Piper over to the side and said, "That is not the way we're supposed to talk or the way we should treat people. We need to be nice."

Piper nodded, wide-eyed and a little bit confused.

The rest of us laughed. Not because Meghan's behavior was actually humorous. Rather, we laughed in the way you sometimes do when you avoid a near accident on the interstate. In a way, that moment showed us what politics might do to someone after marinating in it for way too long. (Mrs. McCain was pregnant with Meghan at the 1984 GOP convention.) Or maybe we laughed because we grew up around politics, but in no way was it a defining aspect of our lives. We had faith, friends, and one another . . . hopefully that would allow us to avoid a fate of being obsessed with a political spotlight.

Of course, it was just September. We were just getting started.

O O O

When it came time for Mom's big speech, we sat in a certain area reserved for the speaker's family. It was located right in front of a bank of cameras. Also, as I looked down, I saw every single famous news reporter that I'd ever seen on television right there. You might think that it would've been intimidating. However, if any of the thousands of cameras filming our every move would've zoomed out, you would've seen all my aunts and uncles, all my mom's aunt and uncles, all our cousins, and my grandparents. They were surrounding us, everyone was having the time of their lives, and it felt like nothing could go wrong when we were enveloped in such love!

Trig sat next to Britta, Willow sat next to Mrs. McCain, Dad sat next to Piper, and I sat between Levi and my grandparents. Around us were dignitaries like Rudy Giuliani and Senator McCain's mother. It was spectacular to be right there, to have a front-row seat to history.

Finally, after so much anticipation, a woman's voice introduced the "Governor of Alaska." When she came out onstage, the crowd erupted. It took over three minutes for her to begin, as she had to wait for the noise level to subside. People were chanting her name, they were holding signs that read Palin Power, and—after a while—Dad looked at us and just started laughing about how amazing this reception was. After a long and divisive battle for Republicans to finally settle on a candidate, I think everyone was just so pleased to have a fresh face onstage. Finally, she spoke into the noise, which made people sit down and start listening. But with her first sentence, the entire hall erupted again.

"Mr. Chairman, delegates, and fellow citizens," she began. "I will be honored to be considered for the nomination for vice president of the United States. . . ."

After the noise quieted down, her speech started out by describing the many qualities of Senator McCain: his military service, his integrity, and his inspirational statement that "he would rather lose an election than see his country lose a war."

Then it was her moment to tell people who she was . . . really. She talked about Track (about to be deployed to Iraq as an infantryman in the United States Army) and my cousin (already in the Persian Gulf in the navy) and how she'd be proud to have Senator McCain leading them as commander in chief.

She talked about the rest of us kids, by saying, "In our family, it's two boys and three girls in between—my strong and kindhearted daughters Bristol, Willow, and Piper."

At the listing of our names, the crowd applauded and the cameras showed us sitting there in a row. The cameramen couldn't seem to find Willow, who was by Mrs. McCain. (They never were really able to tell us apart!) Piper wasn't sure whether she should stand, but finally committed to standing up and waving at everyone. As always, her cuteness melted the hearts of the convention-goers. I smiled, slightly horrified at being stared at by millions of people all at once, and grabbed Levi's hand. Again, I was thankful he was with me.

It was slightly weird to have all the eyes of the convention on us, but soon enough Mom was talking about her "perfectly beautiful baby boy named Trig." She said that if she was elected vice president, families of special-needs children would have a friend—and an advocate—in the White House. The entire audience stood on its feet and cheered raucously. Trig, the subject of the crowd's adoration, remained in sweet oblivious sleep through it all!

Then she talked about the love of her life—the man she'd met in high school. Dad, she said, "is a lifelong commercial fisherman, a production operator in the oil fields of Alaska's North Slope, a proud member of the United Steel Workers' Union, and . . . a world champion snowmachine racer."

When the JumboTron screens showed Dad, the entire convention center—in fact, the entire world!—got to see this "man's man" cradling Trig in his arms. I remember thinking Dad's amazing qualities—both his toughness and his tenderness—were well showcased in that moment. But Mom wasn't content to let her description of him rest there. "We met in high school, and two decades and five children later . . . he's still my guy."

Everyone laughed and cheered at this, and Dad—who had handed Trig off to Piper—stood up and waved enthusiastically to the crowd. They seemed to love him and couldn't get enough of this "normal" family that was already withstanding all the criticism and seemed to be having a lot of fun anyway!

But there was still more family to honor and brag about.

"Among the many things I owe my parents is the simple lesson I've learned, that this is America and every woman can walk through every door of opportunity."

I think, at this point, I could imagine liberals at home watching the television screens and becoming furious. A woman? A Republican? Talking about opportunities? She's stealing our lines! "She" is not supposed to be happening! But the conservative crowd—which knows the Republican Party is pro-women and pro-equality—stood on their feet and cheered as she acknowledged my grandparents. "I'm so proud to be the daughter of Chuck and Sally Heath."

Her speech was about forty-five minutes, and everyone agreed . . . it was a barn burner. About halfway through it, the cameras cut to Piper as she tried to make Trig's hair lie down on his sleepy little head. Of course, the cameras were on us constantly, so we

never knew what was being broadcast to the whole world. She definitely didn't realize that one of the cameras had cut to her and was
broadcasting live as she cradled her new brother and rubbed his
head. His hair wasn't cooperating, even after she tried to fix it, so
she did what any of us would do in the privacy of our own home.
She licked her hand from the middle of her palm to her fingers,
and then rubbed her slobbery hand through his sweet hair. In the
battle between Piper and Trig's hair, Piper won.

It was a very short moment of unscripted "real" behavior, and
America loved it. Perhaps voters were a little sick of the political
bitterness it took to get us to the two conventions or perhaps they
were sick of the overly contrived staging of the two events. Whatever the reason, the video of Piper's "hair lick" became an Internet
sensation.

It may be hard to remember what it was like back on that night
in September. Since she entered the nation's consciousness Mom's
been so mocked, idolized, and mimicked that it might be hard to
remember the moment when you had no idea she was capable of
sentences like:

"I guess a small-town mayor is sort of like a 'community orga-
nizer,' except that you have actual responsibilities."

Or

" . . . though both Senator Obama and Senator Biden have been
going on lately about how they are always, quote, 'fighting for you,'
let us face the matter squarely. There is only one man in this election who has ever really fought for you."

Or the ad-libbed line she blurted out when her teleprompter
stopped working and she spotted a group of delegates in hockey
jerseys:

"Do you know the only difference between a pit bull and a
hockey mom? Lipstick."

The way a convention works is that everything builds to the crescendo of the presidential candidate taking the stage on the last day. Many people wondered how Senator McCain's speech could top my mom's speech, but by that time I no longer was worried. Now that my pregnancy was revealed, my mom had blown the roof off the convention center, and people seemed to genuinely love our family . . . I was having a blast! And I knew the senator would rise to the occasion, as he always did.

After the speech, Mom couldn't walk two feet without people stopping her. People crowded around her like she was a rock star, and she took it all in stride. As I stood back and watched people swarm her, I remember thinking, *From now on, things are different.* I also took a moment to remember the sometimes challenging path she took to get to that spot—how she ran for city council, for mayor, for lieutenant governor (and lost!), and then governor.

It just seemed natural, and humbling, to see that God had taken us down that path as a family. And we were right where we needed to be.

By her side.

By the time the convention was over, it was obvious that something had changed. The beleaguered McCain campaign, which had fought so hard to get to that point, had been revitalized; Republicans all over America were chattering about this newcomer; and our family had come together in a beautiful moment of honor and dignity.

Chapter Nine

Sinking In

W hen we got back home to Wasilla, we were away from the glitz of the convention, the constant pressure to look great, and the television reporters . . . but you can't ever *really* be away from the reporters.

I was sitting on Mom's bed watching Fox News when I heard my name. The host said something like " . . . Governor Sarah Palin confirmed that her seventeen-year-old daughter, Bristol, is five months pregnant and that the child will marry the father."

I'd avoided all of the press when I was with my parents at the convention. Since I was pregnant, I didn't want to cause myself unnecessary stress. But there I was, alone in my parents' bedroom, and I couldn't tear my eyes away from the screen.

They played a clip of then Senator Obama saying candidates' children should be off-limits.

I couldn't believe my pregnancy was being discussed on national television with the other candidates. The screen went back to a "talking head," a political adviser who believed I'd shown courage as my drama sadly played out in such a public glare.

I (a seventeen-year-old kid!) had become a talking point. You may have thought this would've dawned on me sooner, like during the convention when it all became public and cameras recorded our every expression. Though I knew everyone paying attention to American politics was *aware* of my pregnancy, I didn't realize how *important* everyone thought it was. Before the convention, I'd been consumed with trying to figure out what kind of trade I could learn so I could afford to take care of my baby. During the convention, I was so caught up in all the excitement of Mom's speech that it didn't dawn on me how many people were chattering about me. Only when I returned home did it really sink in. And it felt less like a gradual realization and more like a stab.

This was the only time I heard anyone talking about me on television during the election, so I immediately called my mom. After she did her best to comfort me, I sadly hung up the phone and was left sitting in the house alone. That's when I did what any typical teenage girl would do—I called my friends. The only problem was that I was no longer the "typical teenage girl," at least according to the Secret Service that surrounded me constantly.

Our house in Wasilla, which you may have seen on TLC's *Sarah Palin's Alaska,* is right on Lake Lucille. That meant not only were the Secret Service at our front gate and door, they were also sitting in boats all day (and night) on the lake. Plus, they followed us everywhere we went. Once, I was getting my hair done, and my stylist asked in a low, suspicious tone, "Are those people stalking you?" She nodded to two men outside the salon. "They've been out there for a long time."

Because it was Alaska, the Secret Service didn't look like the ones on TV, with dark suits, sunglasses, and sleeves that hid communication devices. Our Secret Service agents wore boots, hats, and parkas to protect them from the weather. This made it even weirder, because it wasn't obvious why they were following us. They did have earpieces, but you'd have to look closely to see them. So it basically looked like we were traveling with our own posse of big burly men. I was somewhat used to having a security detail. When Mom was governor she had security, but not like these guys. The Secret Service followed me everywhere—to school, to the grocery store, and even to Levi's house. Once, I went to Valdez, where he was working—a seven-hour drive—and the entire way I saw their vehicles in my rearview mirror. I bet Mom wished she had that kind of surveillance when I was younger!

Eventually, I'd be glad to have them there. Weird cars crept up our driveway, and reporters tried to sneak in through our bushes. But on that day, as I sat on the bed in tears, the Secret Service had held my friends up at the gate, and I just felt that much more alone.

After a few texts and several conversations convincing security that these girls were actually friends, I had my first big cry of the campaign.

"What am I going to do?" I said, with a tear-streaked face. Of course, they had no idea, and neither did I. Our lives prior to this consisted of snowmachines and trying to get out of math homework, not dealing with the pundits discussing your sex life on national television. There's no guidebook for what to do when thrust into that situation, so we just tried to figure it out as we went along.

And the campaign had barely even started.

Thankfully, I did figure out a short-term plan for life. Since

Mom and Dad were off campaigning, they needed someone to watch the house and take care of the kids while they were gone. I was the perfect girl for the job! This allowed me to live in the house and have a job at the same time. One of Mom's friends also stayed with us.

Though I had taken care of babies my whole life, I was getting useful practice learning how to manage a household. I took Piper to dance classes, made sure I went to all my doctor's appointments, and checked to make sure everyone was doing their homework.

In one eventful moment while Mom and Dad were away, kids at school started making fun of Willow.

"Your brother's retarded!" one kid said to her, laughing.

One of Willow's guy friends beat him up, and he never dared make fun of our precious little brother again. (Other, mean-spirited adults would, but at least this kid didn't.)

Otherwise, my time was spent paying bills and making sandwiches for the girls' school lunches. It wasn't glamorous, but it was a great situation for an unwed mother-to-be trying to figure out a future.

One night, about eleven o'clock, however, my phone started ringing. I knew it wasn't Levi calling from the Slope because we'd already talked that evening.

The number was blocked, which I thought meant it was Mom or Dad.

"Hello?" I asked. My voice was tired, and I didn't try to hide it. Now that they were traveling, it was hard for everyone to remember that we're four hours earlier than the East Coast. I got calls at all times, especially early in the morning.

"Hey slut," the caller said.

"Who's this?"

"Go to hell," he responded, before hanging up the phone.

Now that'll wake you up.

Literally about thirty seconds later, my phone rang again.

More hesitantly, I answered. "Hello?"

"There's no way your mom is going to win this election, so tell her to . . ."

This time, I hung up the phone.

It rang again.

"Skank!" was the first word that came out of that interaction. The weird thing is that they were all different people with different accents. What was happening? I immediately called my mother.

"Mom," I said frantically. "Something's very wrong. I think someone hacked into my e-mail!"

"No, they hacked into mine."

Apparently, someone had gotten into my mom's e-mail account and taken a screen shot of its contents. Then he posted all kinds of personal information on the Internet: private e-mails, photos (including one where Willow was being silly, crossing her eyes and holding Trig), contact information, and phone numbers for all of our friends and family. Though this was illegal, the large media outlets published all of this stolen information. That meant my phone number was broadcast on "legitimate" news stations and blogs all across the nation. (And not only mine, all of our family and almost everyone we had contact with.)

The calls kept coming. All night, every two minutes, someone would call with threats or insults. Track sometimes picked up the phone, furious at the invasion of privacy, and would tell them in very colorful terms to back off his family.

They quickly realized who the culprit was—the son of a Democrat Tennessee lawmaker who said Mom was only upset about the hacking because she was trying to hide something. He later admitted the truth: he was trying to find out something damaging about the family. Something that could hurt the campaign.

Though he found nothing, he created such complications for everyone. It made everything on the campaign trail much more difficult, but it also made everything at home almost unbearable. Immediately, the FBI came and confiscated our phones. To make matters worse, I was under eighteen, so I couldn't legally sign a contract to get a new phone. Someone finally gave me a prepaid phone, but until then I couldn't communicate with Mom and Dad, arrange school pickups, or even give Mom and Dad updates as Track prepared to go to Iraq.

It was so depressing. Levi wasn't even around to help out during this time because he was away most of the time on the Slope. However, we developed a good way to communicate. I gave him a spiral notebook, which he wrote in every night, then when he'd come home, he'd give me a whole lot of letters. They were filled with details of his days, which included twelve-to-fourteen-hour workdays, caribou and fox sightings, as well as little notes to Bentley. It was at least one way to keep connected, though he was very far away from my normal life.

And in the midst of all of this, Track got his orders to deploy to Iraq. After he'd joined the army, he was stationed at Fort Benning, Georgia, and then in Fairbanks at Fort Wainwright as part of the thirty-five-hundred-soldier Stryker Brigade. Mom was the speaker at the deployment ceremony, which she'd committed to in her role as governor long before she realized her own son would be one of the thirty-five hundred.

As Mom spoke, my brother stood at attention amid all of the other soldiers, wearing desert fatigues and black berets, facing the crowd. He completely blended into his unit, nicknamed the "Grey

Wolves," and Mom didn't mention him by name. He never wants to be singled out!

I didn't even get to go to the ceremony. By this time, I was so ashamed at what I'd done, I hated to go out in public and feel everyone's eyes on my ever-growing baby bump. And so I was all alone at home, blinking back tears as my big brother headed out to war.

Now that Track was leaving, who'd teach my son what a real man was supposed to act like?

Of course, nothing could've been a better example of manhood than for my son to know that his uncle Track loved his country enough to fight for it. (This, of course, is a lot easier to write now that he's already back from the war zone.)

At the time, it felt like my whole world was coming apart at the seams.

While Mom was in Alaska for Track's deployment, she came by the house with Charlie Gibson for an interview. The television producers came into our kitchen and made it into a set. The campaign sent Nicolle with Mom, so they went through Mom's closets to select clothes that she could take on the campaign trail. It was one of those slightly awkward moments, when Nicolle scrunched up her face as she was going through each item and said, "No, no, no . . ."

It seemed, once again, that our clothes were not quite up to this vice presidential task.

When everyone left, and the kitchen became our regular old kitchen again, I waved good-bye at the door. Though I'd miss Mom, I was thankful that I wasn't going back on the campaign trail with them.

O O O

On my fifteenth birthday, Mom declared she was running for governor, and on my sixteenth, she was still running for governor. However, my eighteenth birthday really made up for previous birthdays that were less than exciting! Willow, Trig, a sitter for Trig, and I packed up and went to New York City, where my mother was about to tape *Saturday Night Live*.

I was seven months pregnant, and I hadn't seen Mom in a few weeks. When she came into the hotel room with all of her entourage, it was so good to see her. She looked at my pregnant belly and exclaimed, "Oh my goodness, you're getting so big!," and I was so embarrassed.

Mom and Dad gave us cupcakes from Magnolia, a bakery in Rockefeller Center that makes everything from scratch in small batches all day so the food is so fresh. My cupcakes were pink and brown, and I've never had any more delicious! Also, the McCain campaign staff pitched in and gave me a $400 gift card to Target. This was much larger than any of my paychecks, so I was pumped. I dreamed about what kind of cool things I could get for the baby with such a large amount.

I got my hair and makeup done, and then got to go backstage to meet a lot of the *Saturday Night Live* crew. Amy Poehler, also pregnant at the time, compared notes with me about our pregnancies. I also met Alec Baldwin, who stars on *30 Rock*. When it was finally time to go to the taping, I sat in the audience and watched as my mother did the opening sketch.

It began with a mock press conference that had Tina Fey in her now famous role pretending to be Mom. On a different part of the stage, Mom was watching Tina's performance on a television monitor with *Saturday Night Live*'s executive producer.

Mom said, "I just didn't think it was a realistic depiction of the way my press conferences would have gone. Why couldn't we have done the 30 Rock sketch that I wrote?"

"Honestly, not enough people know that show," Lorne Michaels responds. 30 Rock, of course, is the popular show Tina Fey produces.

That's when Alec Baldwin pretended to mistake my mom for Tina-dressed-as-Mom.

He tried to convince the producer not to let his friend appear on the show with her, saying "You want her, our Tina, to go out there and stand there with that horrible woman? What do you have to say for yourself?"

That's when the producer pointed out that Mom was really Mom and not the look-alike Tina.

"Forgive me," Alec said. "I must say this. You're way hotter in person. I mean seriously. I mean, I can't believe they let her play you."

I never thought I'd see my own mother yell, "Live from New York, it's Saturday Night!," but she did. It seemed so surreal hearing those words coming out of her mouth . . . the same mouth that used to nag me about keeping my socks off the floor. The crowd seemed to love her being there, and the SNL staffers backstage were cracking up at some of the skits.

The most hilarious moment came later, during Weekend Update.

Seth Meyers introduced her by saying, "Here to clear up some misconceptions about her campaign is Sarah Palin."

Mom sat behind the fake news desk and said, "Thanks, Seth, but after some thought I've decided not to do the segment we practiced. I've decided it might not be the best for the campaign. It seemed to just cross the line."

Seth turned to Amy Poehler, who sat at the other side of the long news desk, and said, "Amy, would you like to do the governor's bit?"

"Sure," she said. "I guess I could give it a try. . . . I *think* I remember the words."

Then she exploded from her seat and started a rap. It began,

> *My name is Sarah Palin you all know me*
> *Vice Prezzy nominee of the G.O.P!*
> *Gonna need your vote in the next election*
> *Can I get a "what what" from the senior section?*

It was absolutely hilarious. They had guys pretending to be Eskimos and one pretending to be my dad (complete with an Arctic Cat jacket and goatee!) dancing. He was such a huge contrast to my real tough-guy dad that we laughed and laughed.

Then Amy Poehler rapped, "All the mavericks in the house put your hands up! All the mavericks in the house put your hands up!"

When she said this, my mom put her hands in the air and did the "raise the roof" motion. The crowd went crazy.

This is one of the many times I thought, *My mom is the bomb!*

Everyone across America—Democrats and Republicans alike—believed Mom absolutely rocked on *SNL*. Within days, however, people would turn on her once again.

Soon the press would report that my mom was demanding expensive designer clothing and accessories to wear while campaigning.

I was furious. How on earth could people blatantly lie like that?

I know it's hard to believe that there are normal, everyday people in the world of politics, where everyone is so fake. But my mother wears sweatshirts and Carhartts and—unfortunately—

SKECHERS Shape-ups in her normal life (okay, so Ben and I bought her those, but it's still funny). Remember my earlier story? When I was a kid and was complaining at the Governor's Mansion about Mom getting rid of all the luxuries other governors' families had, Mom replied, "We're not like other governors' families, Bristol."

Truer words had never been spoken. Our family wasn't some redneck family that finally got a chance to spend big money on someone else's dime. We've always shopped wisely and bought in bulk at Costco. In fact, when Mom stopped off at that Ohio Walmart, she even made headlines for buying generic diapers for Trig!

None of us ever asked to get nicer clothes. In fact, we would've preferred to wear our normal ones, which looked better and felt more comfortable. Mom was called a "pampered princess," and public messages were sent for her to return her nylons! Talk about low class. Worse, no one came to defend us, so Mom decided to set the record straight herself at a Tampa rally.

"Those clothes, they are not my property," she said. "Just like the lighting and staging and like everything else the RNC purchases. I'm not taking them with me. And today I'm wearing my own clothes from my favorite consignment shop in Anchorage."

And that was literally true. She had on a Dolce & Gabbana jacket she'd bought there, along with a pair of Paige jeans. (Paige Adams Geller is a Wasilla girl who's now a big L.A. fashion designer!)

Ever hear of someone getting thrown under the bus? It's even worse when your family is thrown under the *campaign* bus.

My grandma is very civic-minded and has worked the polling stations for as long as I can remember. Whenever Mom is a candidate, however, Grandma steps aside to make sure that there's no appearance of wrongdoing. The polling areas have strict restrictions about politicking within a certain amount of distance from the voting booths. This means that you can't carry signs, wear T-shirts, or even talk about your candidate within two hundred feet of the polls.

I'd never voted before, so I was excited to be able to cast my first vote for my mother for vice president of the United States. Grandma was so proud as we drove up to the State Elections Building. We were voting early so we could go to Arizona for the actual election night.

Grandma was about to burst, because you can't talk to people once you get into the building, and she wanted to talk about how she was voting for her daughter. When we walked in, it was all business, except Grandma beamed from ear to ear.

The other people in the room noticed her—she's well known and well liked in town—and there were a lot of winks and silent thumbs-up. It was a very proud moment for both of us.

A few days later, we were traveling all the way to Phoenix, where the GOP was having its victory party. That's what they always call the election night results gatherings, no matter how unlikely victory appears to be!

Once again, everyone made the trip: Dad's parents from Dillingham, Grandpa and Grandma, Aunt Heather, Aunt Molly, Uncle Chuck (and all of their families); my aunts and uncles on my dad's side; Dad's Iron Dog partner, Scott Davis, and his wife, Kris Perry, and her family; Meghan Stapleton (who'd been instrumental in helping to hold down the fort during the campaign); and of course all of the Palin kids except Track, who was in Iraq. We even brought some friends! (Levi wasn't able to come, though, because this happened while he still had a job.)

It's hard to understand how far away Alaska is, but these guys went to great effort and cost to get down to Phoenix. Some of them were from places in Alaska where you have to take a "puddle jumper" to get to Anchorage. Then it's usually one or two flights to get all the way to Arizona. Needless to say, after such a whirlwind, it was good to be around people who really loved our family.

It's probably unnecessary to write this, but we lost. Even though it was on a much larger scale, I felt the same way about her loss of the vice presidential race as I did about her losing the lieutenant governor race back in Alaska. The entire process was a long shot, God was in control of the outcome, and I was proud of her for trying.

Though it would've been absolutely amazing to win—and we all wanted a victory so badly—my family has always just put things in God's hands. When John McCain gave his concession speech, I was in the audience trying to hide from view. By this time, I was so far along that I could barely travel. I definitely didn't want to be in the national spotlight. But as I listened to Senator McCain's gracious speech, I was moved by what our family had gone through in the past few months. Especially when he thanked my mother.

"I'm also, of course, very thankful to Governor Sarah Palin, one of the best campaigners I've ever seen and an impressive new voice in our party for reform and the principles that have always been our greatest strength. I thank her husband, Todd, and their five beautiful children for their tireless dedication to our cause, and the courage and grace they showed in the rough-and-tumble of a presidential campaign. We can all look forward with great interest to her future service to Alaska, the Republican Party, and our country."

Maybe it was just me, but when he thanked my mom it got the most applause of his entire speech. Mom stood behind him next to Dad with tears in her eyes.

The day after the election, Dad delayed our departure so we could soak up a little more of the Arizona sun. It was November, and we weren't used to this wonderfully warm weather. We were at the Arizona Biltmore, so we sat outside by the luxurious pool, which was very relaxing. The palm trees hung over the pool, and I took a long look at them. Not many of those back in Wasilla! Mom dipped Trig's toes in the water, Willow and I sat on the beach chairs, and—for the first time in weeks—everything was suddenly calm.

It wouldn't last long.

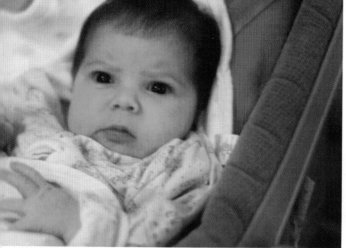

I was the first granddaughter born on my mom's side, so all of the aunts were thrilled to have a chance to spoil me. I was born on October 18, 1990, at Valley Hospital in Palmer, Alaska.

My middle name is Sheeran, after my maternal grandmother's father, Grandpa Sheeran.

This is the first surprise Aunt Molly gave me—a dog I named Indy. There would later be two other dogs named Indy!

When I was three and Track was five, we helped my grandpa work on his antler collection—it is now three times the size.

Holding my baby cousin Karcher in 1995.

Willow (two), Lauden (three), and me (six) at the Bruces'. My parents had just given us "pleather pants," and I wore mine every day.

With Willow on a family vacation in Hawaii. The sunshine was a welcome break from Alaska's weather.

Aunt Molly enjoyed dressing Lauden and me in matching outfits.

Willow (two), Lauden (three), and me (six) at the Bruces'.

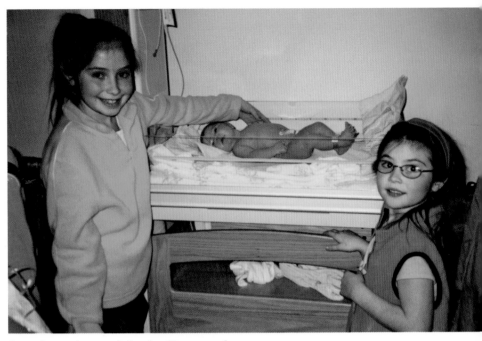

I was beyond excited the day Piper was born.

I was thrilled to have a real-life baby to play with instead of playing pretend with my dolls. She definitely got a lot of attention!

I fished in the Mat-Su
Borough from a young age.
Here I am fishing with my
grandfather at Willow Creek.

After a summer basketball camp with Coach Bradley, I enjoyed hanging out with my sisters.

Piper and my cousin McKinley, both one, while I watched them at my grandma's house.

I spent my school breaks attending Coach Bradley's basketball camp.

I joined the football team when Track insinuated that I couldn't handle the practices. I had to prove to him that I could!

Posing next to fossils at my grandparents' house. These same fossils were later inlaid in my parents' fireplace.

The cousins gathered at my grandparents' house during Christmas 2004. This is a family tradition that we do every year.

The entire family lining up for a snowmachine ride. (*Courtesy of Sarah Palin*)

A photo spanning the generations, including my great-grandmother, grandmother, and sisters, celebrating our Alaskan heritage by wearing Kuspuks. *(Courtesy of Sarah Palin)*

When I was a freshman, I went to a school dance with a friend.

After my mother became governor, we posed with the newspaper the next morning. (*Courtesy of Chuck Heath*)

The governor's gala helped celebrate my mother's gubernatorial success. She was the first woman elected governor of Alaska. (*Courtesy of Chuck Heath*)

I was thirty-two weeks pregnant
when I posed for this photo.
(*Michele Ireys Photography*)

Tripp was born on December 27, 2008.
Levi went with me to Tripp's first
doctor's appointment on December 31.
A few days later we had already
broken up.

Levi and I went to a wedding bazaar to help plan our
wedding, and we jumped into a photo booth. Tripp
was just a few days old. Only days after this photo was
taken our relationship was over.

I was so excited to bring Tripp home from the hospital.
My family turned part of my childhood bedroom into
a nursery that I shared with Tripp. Having a newborn
was a lot of work, but there were so many wonderful
moments like this one. I am so blessed to have him in
my life.

I was the only student to accept a diploma while wearing baby puke on my dress. Piper is dressed up in my graduation gown while I hold Tripp and McKinley looks on.

July 2009: I love the cute dimple above Tripp's cheek.

Summer 2009: I dressed Tripp up like the little man he is.

My dad taught me the invaluable skill of swaddling a baby.

Because Tripp and Trig are so close in age, they love playing with each other.

Track hanging out with Tripp at my house in Anchorage. Track is a good example of what a real man should be.

My mother gives Tripp a kiss in front of my truck, which has a license plate that shows exactly where my heart is.

My cousin Kandice, my mom, me, Track's fiancée, Britta, and Piper went to New York City for the 100 Most Influential People party.

On the first night of *Dancing with the Stars,* I wore a conservative gray suit jacket with an American flag pin on my lapel. But just a few seconds into the dance, I ripped off the costume to reveal a red fringed minidress underneath. Mark and I earned a combined score of 18, which wasn't too bad for a girl from Wasilla who'd never danced in her life. (© *Adam Larkey / ABC via Getty Images*)

One of the most challenging parts of competing on *Dancing with the Stars* was not being able to hang out with Tripp as much as I wanted to! His on-set visits energized me.

After the glamour, costumes, and craziness of *Dancing with the Stars,* it was humbling and eye-opening to travel to Haiti with Samaritan's Purse. Meeting children like these memorable ones really tugged at my heartstrings and gave me a healthy dose of perspective. (*Courtesy of Sarah Palin*)

Our home overlooks Lake Lucille, which freezes completely solid during the winter. Tripp is learning how to ice-skate on this cold day. Maybe one day he'll follow in my brother Track's steps and become a great hockey player!

I can't believe how quickly Tripp is growing up. I'm proud that he's becoming one cool kid, and I can't wait for what the future holds for him.

Chapter Ten

Not Picture Perfect

I f that b—ch comes over here, I'm going to kick her ass," Levi's sister Sadie screamed into the phone.

Levi had just told his sister we wanted to stop by her house to drop off Christmas presents for his family. I looked down at my pregnant belly and just laughed.

"Really?" I said in disbelief to Levi, who had just hung up the phone. "Your sister is going to beat me up while I'm carrying *her nephew*?"

Sadie had always been unusually close to her brother, which translated into jealousy toward me. For example, she was jealous when Levi got my name tattooed on his ring finger, and once went to Mexico for vacation and came back with a tattoo of his name on the inside of her wrist. It had hearts and angels' wings around the word *Levi,* and she was only about fifteen or sixteen. On her other wrist, she has a tattoo of the word *Nonny,* which is what she calls her mother, with dolphins surrounding it. Once Levi and I got engaged, she referred to my mother as "mommy-in-law" and to Trig as her "baby brother."

Sadie never liked me and always seemed jealous of my relationship with Levi. Good friends with Lanesia (who, you might recall, chased my friends and me around the parking lot trying to beat us up), she'd frequently invite her friend over to spend the night, seemingly to get Levi interested in her friend.

"What is up with them?" people sometimes would ask me, with suspicious looks on their faces.

I just shrugged.

But I didn't have time to think of that now. I had a stack of Christmas presents to deliver, and I wanted to get them to Levi's house before I delivered that baby. "Are you gonna let your sister talk about me that way?"

"I can't do anything about it," he said.

Throughout my pregnancy, things between Levi and me had been pretty stable. He had a good job on the North Slope, he seemed to be dedicated to making things work out, and we were still on the "after campaign" high of having had a good, meaningful time together at the convention.

"I think I've changed my mind about the name," I said. Though I loved the name Bentley, I was thinking of going with a *T* name, to go with Todd, Track, and Trig.

"What's wrong with Bentley?" he asked. We'd been talking about "baby Bentley" for so long that it seemed weird to change it.

"I was thinking about . . . Tripp!"

Levi liked it, but he had already become emotionally invested in Bentley. In the end, though, "Tripp" won out. We also decided on middle names. Our friend Ben suggested Easton, a name I love because it's a hockey brand! Also, I chose Mitchell, because it's my dad's middle name.

Even as the relationship between Levi and me was seemingly improving, though, my relationship with the Johnston family was going downhill and fast.

On the Thursday before my baby was due on December 14, Levi texted me.

> **You aren't going to believe this. Guess who just raided my house and destroyed everything? My house is like a tornado. They think my mom sells drugs!**

By this time, I'd already accepted some of Levi's limitations, one of them being his loose grasp of the truth.

But that evening, it was no joking matter. The news confirmed what he had told me—though I doubt they "destroyed" his house—was actually true. The troopers had apparently been conducting an undercover drug investigation since the previous September and charged Levi's mom, Sherry, with *six* felony counts of misconduct involving a controlled substance.

It started to all make sense. Maybe that's where he got the money to buy me Coach bags and nice rings. I never really thought about where his mother got her money for her tanning bed visits, acrylic nails, or new cars with rims on them—probably from selling her prescription meds!

Apparently, the police had informants who bought OxyContin from her. The next day, she sold them the drugs again, and then she did it again in November. That last drug deal, in the Target parking lot in Wasilla, was even videotaped. Needless to say, based on the advice of her attorney Rex Butler, she pleaded guilty.

As I approached my due date, I was really trying to get rid of all the stress in my life, and this news didn't help. How weird was it that one of my baby's grandmothers would spend time in jail for dealing drugs, while the other might spend time in the Oval Office?

I tried to push this new information out of my head as I prepared for Tripp to arrive.

My due date of December 14 came and went. That was actually Levi's mom's birthday, and Sadie's birthday was one week later.

"Well, you better not have Tripp on my birthday," she told Levi and me impatiently. "Because that's *my* day!"

Okay, I thought sarcastically. *If I start having labor pains on that day, I'll try to cross my legs and hold out until after midnight so I won't steal your thunder.*

Day after day passed, and it didn't seem like we'd progressed at all.

During the last week of my pregnancy, I had to go to the doctor every day, because Tripp didn't seem like he was interested in moving. I'd trudge to the appointments by myself, and sometimes Grandma would take me. (When I was tested for gestational diabetes, Grandma took me for all those tests too.)

Other than the near-constant appointments, I felt enormous and just stayed home as much as possible. Even some of my best friends—like Ben—didn't see me when I was big at the end of my pregnancy.

On December 24, we decided to go to a Christmas Eve service. Track's long-term girlfriend Britta's father is a pastor at a Lutheran church in Wasilla. So I hauled my very pregnant body into the church and barely paid attention to the service. How could I? With every passing minute, I kept thinking that I could go into labor at any moment.

Finally, the doctors told me that if I didn't go into labor, they were going to induce me.

Thankfully, they didn't have to. On Friday I was at home with Mom, and my back was hurting so badly I couldn't bear it anymore.

"Mom, if this isn't labor, I don't know what is!"

She packed up Trig, and the three of us drove to the hospital. Dad and Levi were at the cabin snowmachining—part of Dad's continued effort to invest in what he thought would be a lifelong relationship. When Levi got the message that I was actually in labor, he immediately came to the hospital.

I didn't have the easiest labor. The cord was wrapped around the baby's neck three times, and it took a lot to get the little thing out into this world. But finally, two weeks past his due date, Tripp Easton Mitchell arrived on Saturday, December 27, at 5:30 in the morning, a healthy bouncing baby boy! He weighed seven pounds, three ounces. The doctors tried to hand Levi the scissors to "cut the cord," but he backed away. He said it was too gross, an odd statement since he's able to field dress a moose with one arm tied behind his back. Mom graciously took the scissors and performed the task for her grandson. It was the first time my parents had to step in and do something that the father really should've done, but it was far from the last.

The experience, of course, was the most amazing, life-defining moment for me. When the doctor laid Tripp in my arms, I knew this baby was not a mistake. Having sex outside of marriage was the mistake. But this baby? He was—and is—a blessing.

This very simple concept baffles the media. It was like we had two options: to say that premarital sex was morally acceptable and that Tripp was a wonderful new addition to the family, or to say that premarital sex was morally wrong and that having Tripp was a total disappointment. They seemed unable to understand that sometimes the most amazing blessings can emerge from your worst decisions. But isn't that frequently the case? Don't you see that in your own life? The Bible says, "And we know that all things work together for good to them that love God, to them who are the called according to His purpose." I like the way the King James Bible chooses to start that with: "and we know . . ." It seems to point out how obvious the idea is.

Yet the media was always trying to define me as a hypocrite for having Tripp.

Think about that for just one second. When a professional athlete or Hollywood actor goes to rehab, it is considered "news." When a famous person goes in and out of a clinic, over and over, we are encouraged when the person finally gets it all together. When he or she hits the talk-show circuit, it draws huge ratings. The individual is hailed as a hero. When a lifetime gang member leaves gang life and goes straight, vowing never to return, we applaud his or her courage and commitment. We put those people on TV, in movies, and across the airwaves so they can share their journey . . . hopefully to inspire others in another direction. We read their stories with hope that life can be different, for everyone. When people walk away from the bottle or pills or the needle, we celebrate their escape from a life of bondage and pain. They make a public commitment to never be involved again. We send them to schools all over the country so they can show their needle-scarred arms, telling stories of their hallucinations and depression, all in hopes of discouraging drug use. We are inspired by their stories of determination to never go back, even for a moment, to a life of addiction. These people are seen as great examples of courage. Heroes. Role models.

In fact, some of the best spokespeople, for any cause, are the people who have experienced the "other side" . . . and lived to tell about it. But when a twenty-year-old single mom like me wants to encourage people to wait for the appropriate time for sexual activity, the rules change. The media creates stories that don't exist and passes them off as "news" or "truth." Comedians attack me. "Journalists" malign me. For some reason, I'm seen as fair game.

You see, in the media, I'm not called a role model. They call me a hypocrite for wanting to encourage people to save sex for the marriage bed. People picket, protest, attack, and malign me in all forms of media. But why?

I don't think it's because they don't like me. Truth is, they don't even know me. I think if they could come into my home and we had a chance to talk over a cup of coffee, we'd probably hit it off.

So why am I labeled a hypocrite? Because the precious child I was holding at that moment in the hospital showed the beauty of life, even in less-than-ideal circumstances. When my family and I see beautiful Trig with his unmistakable Down syndrome characteristics, we see God's handprint. Many other people see a baby they believe shouldn't have been born. (And, as I mentioned, the vast majority of Down syndrome babies are aborted before they have a chance to draw breath.) So our very struggle through these issues offends critics. It causes them to think about the scars—emotionally and socially—that happen when people become sexually active. They don't want to see the babies that are created or come to terms with the number of babies never born because the mother has made a horrific choice to end their lives.

As I sat there in the hospital with Mom on one side and Levi on the other, it was the picture of "less-than-ideal circumstances." Levi would later tell his friends that being in the delivery room with me was "gross." After I'd wasted more than two years of my life in that relationship, he'd run off with another. And then another.

Sadie came later that night. Though I didn't want her to come—her threat to kick my ass a week ago still stung—I also didn't want to keep her from seeing her nephew. I did, however, make sure my dad was there.

She held Tripp and stayed in the room maybe for five minutes before handing him back.

"My ride's waiting," she said, "so I gotta go."

Again, it was a less-than-ideal circumstance, because people's anticipation was tempered by the fact that I was so young and unmarried. So I didn't tell anyone that I was in labor until after I'd delivered, until Tripp had already arrived. I didn't send out a "mass text" to all the contacts in my phone, and I certainly didn't post on Facebook that I'd delivered. Imagine how surprised everyone was when they showed up at the hospital thinking I was in labor and I already had the baby in my arms.

With everyone who visited, however, I said, "Don't take any pictures! We're only taking pictures with my camera."

That might sound like a weird request from a proud mother.

However, as soon as it was confirmed that Levi and I were expecting a baby, the celebrity weekly magazines started a bidding war for the first photo. It might sound crass, but it's a very common practice. The famous person negotiates the exact terms about how and when the baby is revealed. When the magazine publishes the photo, it drives sometimes millions of readers to either click through to their website or buy their magazines off the newsstands, so it benefits the magazine and the new arrival. Sometimes the celebrities donate the money to charity or sometimes they set up a college fund.

When I was first approached by representatives of a few magazines that expressed interest in the photo, they offered me so much money that it got my attention. After all, I wasn't going to be like Michael Jackson and put a blanket over my child's head in public. I planned on living a normal life with my sweet child. And whoever snapped a photo could sell it for hundreds of thousands of dollars, or a magazine could send a reporter to stalk us and get one for free.

I decided to take the offer with a reputable magazine for Tripp's first photo. Levi and I were so excited that the inevitable magazine photo would at least benefit Tripp and set him up for a very comfortable life. It was more than an apprentice electrician and a coffee barista could've made in decades. It could have bought a nice house for him to grow up in. (Or two nice houses.)

That's why I was so serious about no one taking photos and e-mailing them to friends. There would be time for showing off his sweet blond curly hair and his shockingly blue eyes. But now was a time to be discreet as we prepared to properly announce him to the world. I snapped candid photos on my phone to put in his baby book. I printed them out and handed them to Levi, with a stern warning: "Remember, don't let anyone see them yet." On the back, I wrote "Do not sell or distribute" just as a reminder. The first photo would appear in March!

People kindly obliged in not taking photos, other than with my camera, and Dad and Mom were constant companions at the hospital. Eventually they had to leave for a while to run errands. When they did, Levi came in the door.

I was filling out the birth certificate papers by myself.

"What are you doing?"

"I'm filling out Tripp's birth certificate."

"What are you putting his last name as?"

"Until we get married, I'm putting it as Palin."

"Don't be such a b—ch, Bristol. I want my name on it," he demanded.

I had no idea what to do. Dad had warned me not to allow it, since he didn't trust Levi as far as he could throw him. But Levi was pretty adamant that I hand over the birth certificate. I had two choices. I could argue with Levi as I lay in my hospital bed or I could give in. I was tired, I was hopeful that Levi would magically turn into a good father, and I honestly—after twenty-four hours of labor—didn't have any more fight left in me. I filled it out and—against my better judgment and the advice of my dad—put Levi's last name on there. I've regretted it ever since.

Levi did stay with me while I was in the hospital, and he slept in a chair. The weird thing was, he wouldn't change Tripp's diapers, he wouldn't hold him unless I asked him, and he wouldn't help me get around. I got the feeling that he was there because he was supposed to be there, but he was trying not to be sickened by all that was going on around him. When it was time for me to go home and start trying to live with the baby without the help of nurses and lactation consultants, I was a little worried. Levi stayed at our house during the first few nights. And he was even less helpful. I don't blame him for not wanting to change a dirty diaper—no one wants to do that—but I did start wondering why he was hanging around if he was going to be causing me more work instead of less.

On one of these first nights home, Levi and I were sitting on the couch. I'd been taking care of Tripp and was completely exhausted. Have you ever heard the phrase "sleep like a baby"? Well, apparently Tripp hadn't. The first few nights were just so terrible as I tried to figure out how to breast-feed and sleep and brush my teeth all in the same day! I'd just changed Tripp's diaper for the millionth time and put him down to sleep when Levi's phone went off. I saw him read the text and put the phone back on my end table.

When he got up to go to the bathroom, I noticed he'd left his phone.

Now, let me say that I'd never snooped. It wasn't like me to be suspicious. (In fact, I wish I had been more suspicious before this point!) But I hadn't slept in days, my patience had been worn thin, and there his phone sat, begging me to pick it up to see if the nagging feelings in my heart were paranoid impulses or truthful nudgings.

Casually, I picked it up—it was the one I'd bought Levi for Christmas. He probably didn't think too much about leaving his phone lying around, because I wasn't the type of girl who snooped. I wish I could tell you that I looked deep into his archived texts and e-mails, and hidden amid a ton of information, I finally came across something that incriminated him. But it was right there. At the top. It took me less than fifteen seconds to find and read a certain text.

He came back from the bathroom and said, "What are you

doing?" He tried to grab the phone. But by the time he lunged for it, I'd already thrown it against the wall, breaking it.

Because I'd had a chance to read it.

It said:

She had fun and wants to hook up again with you.

"Levi, what was that?"

"Just my sister. She was telling me that my friend was going to buy me chew." He was trying to make me believe the text was about a hockey player friend instead of a girl.

"Your mom's bought your chew since seventh grade," I said. "Why would you have someone else do it now? And why did your sister call him a 'she'?"

He denied it and he denied it, but eventually even he realized there was no use.

"I know you're cheating on me, and I'm not dealing with it anymore! Why would I put up with that, when you won't even help me change a diaper?"

I ran to the balcony where Dad was rocking Trig to sleep, and yelled, "Dad, guess who cheated on me!?"

Dad looked up and just shook his head. He knew what was going on.

I turned to Levi and waited for an explanation.

He no longer tried to defend himself. Instead, all that happened was that he got up and simply walked out the door.

It took that stupid text message to *finally* get my attention. My son was not going to grow up thinking that Levi is the way men are supposed to be. After all, Tripp did not deserve to have such a bad example. So when Levi got up and walked out of the house, I promised to never open it back up for him.

O O O

Right after Christmas, Mom had packed her bags and headed to Juneau, because she was still governor. So she, Dad, Willow, Piper, and Trig headed to the capital without me. I still had key advantages many teen moms don't have. Number one, my parents paid my medical bills that health insurance didn't cover. And number two, they let me live at their house rent-free.

Because Tripp was born over the holiday break, I didn't miss much high school at all. Dad had watched him some mornings so I could take a shower and get ready for school. When he couldn't watch him, I'd scramble to find someone else. (Aunt Molly would watch him early in the mornings sometimes before her own shift at work, because we knew Levi's family wasn't an option.) It wasn't easy, but I pressed through. And not once did Levi volunteer to help. (Or respond to my requests for him to help.)

The first day I went back to Wasilla High School after having the baby, I thought back to how scared I was as a freshman. I thought of Track blasting his radio and laughing at my nerves. It seemed so carefree and silly, compared to my current life. Or the walk of shame I was about to make.

On the way to school that morning—and many mornings af-ter—I had to endure the pain of seeing Levi's truck at the school because his new girlfriend was a younger student, and he'd hang out in the parking lot acting immature with his friends. That morning, I opened the door and forced myself to enter. There's something not right about a high school kid being a mother. It seemed like I'd aged forty years since I was last there. My mother always said I had an "old soul," but now I seemed totally different from all of my girlfriends. Though they were nice to me about Tripp, conversation always went back to boys and "can you believe she's wearing those jeans?" It was hard for me to identify with them, and gradually the conversational distance between Tripp's crib and their shopping sprees got too large to cross.

On an afternoon after I went back to school, I was sitting in the kitchen feeding Tripp.

"Hello?" I heard come from somewhere in the house.

My pulse quickened.

"Who's in this house?" I yelled. I had no idea who would just barge right into our house, and I regretted that I hadn't locked the front door.

I turned and saw a soldier standing right there in our living room.

Track.

"What are you doing here?" I screamed, instantly grinning from ear to ear. I'd never seen him in his uniform, and he'd come home from Iraq for his leave. Typical Track, he didn't tell *anyone* when exactly he'd arrive home. His best friend picked him up from the airport, and he simply showed up. After we visited, he drove into Anchorage and surprised Mom and Britta.

"How could you surprise us this way?" Mom asked, obviously thrilled and emotional at seeing her son. He'd been gone a year, and it seemed that he had grown up a lot.

There was a lot of that "growing up" going around.

I was happy that Levi was working so hard on the Slope and

trying to prepare a better life for our new "almost family." When we got together, we exchanged our handwritten notes. They warmed my heart, though now when I read them back, I'm astonished that I wasn't bothered by the fact that he called me "princess," but spelled it "princes." And "I no I'll see you soon," instead of "I know I'll see you soon." Love has a way of making even problems like that seem endearing.

Except there was one complication . . . being an apprentice electrician is a highly valued position, and competition is fierce for them . . . and one of the qualifications is a high school degree.

One day an Anchorage radio talk-show host talked about Palin gossip in the news once again. But now Levi was the center of a lot of interest as well. After Tripp was born, all kinds of news programs talked freely about our circumstances, that Levi was a high school dropout and I hadn't yet graduated. It raised some questions. Like how he could be in the apprenticeship program without a high school diploma.

Instead of assuming what any friend of Levi's could tell you— that he lied on his application—the radio host raised an ethical question about my mom. Did she use her connections to get her grandson's dad a good job?

Mom was shocked.

"Why did he sit around our kitchen table telling everyone that he passed his GED if he hadn't?" she said when she found out.

It was just another slap in the face. After running as an ethics reformer for governor, she was getting her name dragged through the mud again. Eventually, Levi's dad admitted his son wasn't qual- ified for the job, that he'd gotten Levi the job, and my mom had nothing to do with it. He then went on to say that he and Levi had talked about it and decided for him to focus more on his education. (That was a nice way to spin his lying on a job application.)

This news didn't upset me as much as you might think. Since Tripp's birth Levi hadn't provided for us financially at all. Though his mother had bought Tripp a couple of baby outfits when she found out I was pregnant, my mom and dad ate the cost of the hospital bills.

The public humiliation of Levi losing his job was just further proof that I'd made the right decision in closing that chapter of my life. So I tried to forget him, concentrated on my studies, and tried desperately to escape Levi's fate by focusing on graduating from high school myself. It was actually hard to concentrate on studies.

On March 11, 2009, I was sitting at home, trying to get Tripp to eat, when I found out that *Star* magazine had published an article under the headline "World Exclusive: Bristol Palin's Bitter Split!"

This was the first time that people who didn't live in Wasilla knew that Levi and I were not headed toward marriage. I'd not publicly announced it, because . . . well, it was humiliating. Not only was I a teen mom, I was a teen mom without a marriage prospect. The article also included Tripp's first baby picture, published for the first time without my permission.

When I saw the magazine, I saw right there on its pages a photo I had taken and let Levi have from my camera. Who had the photo? Who knew the story of our breakup? Only Levi and Sadie. I never would've thought that giving my son's father a photo of him would end up costing me so much money, but it did. The *Star* magazine article violated my arrangement with a more reputable magazine for exclusive rights to Tripp's first photo.

And rumor has it that *Star* paid only $5,000.

I was a combination of furious and heartbroken. How was I going to provide for Tripp for the next eighteen years? That one photo could've provided everything we needed: a home, a college education for both of us, a stable life. But in one moment, it was completely gone. I felt such a sense of loss—not just a loss of money, but a loss of power over my circumstances. I'd wanted to go to school, maybe go to college to get my nursing degree, and then start the rest of my life with a husband (first) and a baby (second). Now, I was so limited in my options that it seemed impossible I'd ever be able to make enough to even justify working, considering the cost of babysitting. As I looked at that magazine, it just symbolized so much all at once: my public breakup, my way-too-dramatic life, my awful relationship with Levi's family, and my financial hopelessness.

That week, I hit rock bottom. On top of all the other things going on, I hadn't slept in so many weeks. (No wonder militaries use sleep deprivation to get information from enemy troops . . . it felt like torture to me!) One night, I was up yet again with Tripp, and—instead of falling back into bed—I sank down in Dad's black recliner in our living room, in the big empty house. *Life sucks,* I thought, *and I can't figure out how to live it anymore.*

It may sound odd, but I hadn't really prayed about all of this yet. While my mother had a definite moment of conversion, which was followed by her baptism near Big Lake, I never had that memorable moment on which I could look back and say, "This is when I gave my life to Jesus." I always just knew God was a part of my life, but He was more of an idea, in the background. Even when I was at the Christmas Eve service before I had Tripp, I didn't pray about the delivery or about being a mom. I simply was so consumed with life that eternal things got pushed out of my mind by the more pressing "have to do" things.

But on that night, after so many weeks—so many years actually—I finally called out to Him.

"Help me, God," I said, with a weak voice. "I'm broken. Please fix me."

So it wasn't a big, long, drawn-out prayer with fancy religious language. It was just a plea from the bottom of my being. Tears ran down my face uncontrollably. I was so desperate for someone to help me get through not only Tripp's infancy but also the rest of my life.

"The public breakup, the taunting in the media, the isolation I feel at school . . . I'm done being in denial about all this. I need light at the end of the tunnel, I need you to help me . . . to rescue me."

I prayed about everything—for Tripp, for the Levi situation, and even for opportunities to get me out of the rut I felt stuck in. I needed something to uplift me. "Just help me get through this."

That moment in the recliner was a turning point for me, because God's forgiveness is frankly a very good deal. Not only does He absorb some of the debt of my sinful decisions (in Jesus Christ), He doesn't require me to repay the debt through my good decisions or efforts. He no longer sees my sinful self (lying, drinking, sexually unwise); He only sees me for who I am in Jesus (redeemed, forgiven, and pure). That means I don't have to carry the shame of sins committed in my past with me forever.

To be sure, I didn't get up from that chair and—suddenly—life was fixed. I still had to deal with all of my problems, which would actually even get much worse. But the difference wasn't between a troubled life and then a miraculously carefree life. Rather it was the difference between struggling against my problems alone and with Christ who loves me and has forgiven me.

When I got up from the recliner, I felt lighter and more hopeful. I'd need that extra strength soon enough.

Around this time, apparently, Levi had gotten hooked up with some "handlers." Handlers, for those of you who don't live in California, are people who try to help you milk all you can out of your temporary fame. Like agents, but not as choosy. One was Rex Butler, the same attorney who helped Levi's mom with her drug charges. (Yes, I realize as I write that sentence that my life had become a *Jerry Springer* episode, and a bad one at that.) Rex represents people charged with drive-by shootings, homicides, personal injury/auto accidents, and drug dealing. The motto on his website is "Playing defense, the law and you." He also is a black Democrat and told the press that my mother's policies on African American issues could be summed up in this phrase: "Don't need them, don't worry about them."

Rex's partner, Sherman Jones, is an enormous, meticulously dressed African American whom you could totally imagine as a bodyguard or a nightclub owner. (He's also known as Tank.) He wears pinstripe suits, large watches, hoop earrings when he's dressing up, and velour sweatsuits when he dresses down. His website reads: "Who I'd like to meet? You . . . If you believe that your signigicant [*sic*] other is having a fling . . . We'll unleash the truth."

Rex and Tank—and suddenly Levi—wore Bluetooth headsets and carried BlackBerrys. They started calling themselves "managers" of his career, our hometown "the 'Silla," and Levi "Ricky Hollywood." In fact, once I called his phone and it went straight to voice mail, which said, "Riiiiiicky Hollywood is not available right now." Rex, Tank, and Levi became inseparable.

Levi usually wears flannel shirts, boots, and Carhartts with a ring on the pocket from his Copenhagen tobacco. He hunts and fishes, but could barely carry on a conversation in Wasilla. But Rex and Tank convinced him to be more exciting when talking to the media. Even though the first suit he ever wore was the one he was loaned at the GOP convention, he began wearing sweater vests, skinny jeans, deep Vs, and sunglasses.

While I was taking Tripp to doctor's checkups, trying to arrange for child care, and breast-pumping in the car before school, Levi was scheming with these guys about how to take advantage of his new opportunities. Tank famously said to the *Anchorage Daily News,* "So now what do he do? Go work at McDonald's? So people can ask him, as he makes $8 an hour, 'Hey, how's Sarah Palin? How's Bristol?'"

No, Levi, Rex, and Tank had their eyes set on the money of fame, and they set out on a media tour. Imagine how embarrassing it was to go to school while Levi is talking about your sex life on national TV!

"Hey, did you see the Tyra Banks show?" a friend in the hall asked me in April. He was referring to the fact that Levi and his sister went on that show, on which Tyra Banks asked him if we practiced safe sex.

Though I tried to stay away from all of that, I found out all the details. Levi told Tyra Banks that we did, in fact, practice safe sex. Then she pressed him, since a child had resulted from our sex lives. "Well, what happened? Did you have a wardrobe malfunction?"

"I guess."

"Every time, you practiced safe sex?" she asked.

In the background, his sister starts shaking her head no. (I think if a sister knows about her brother's frequency of contraception use, the relationship is way too close!)

"Every time?" Tyra asked, with disbelief in her voice.

"Most of the time," he said.

"There ya go!" she exclaimed like she'd won at the slot machines. Everyone in the audience laughed.

This was my life. Online courses at home, balancing a baby on my hip, while the most intimate details of my life were shared on national television. Almost everything Levi said was false, but the facts didn't stop him. Later in the year, he posed for a *Vanity Fair* piece, posed completely nude in *Playgirl* magazine, and starred in an extremely stupid pistachio commercial.

Yes, a pistachio commercial, along with Tank. Levi's commercial had his "bodyguard" making sure the coast is clear before he breaks open a pistachio. Then a voice-over comes on and says—wink, wink—"Now Levi Johnston does it with protection."

In the meantime, I got a chance to use my situation to warn other teen girls about having sex before they're married. The Candie's Foundation contacted me and asked me to become one of their ambassadors. It was an opportunity for me to talk about an issue that I realized all too well had real-life, far-reaching implications. If I could help one teenager avoid a similar situation, then it would all be worth it. And that's how I joined the discussion about one of the most important issues in our nation.

Many people didn't like it. In fact, those on both ends of the political spectrum objected to my new Candie's gig, saying I was a hypocrite giving "unrealistic" advice.

My mom got irritated at all of the criticism—she is a Momma

Grizzly, after all! In her book *Going Rogue,* she summed it up nicely by writing, "Bristol isn't trying to write a Sex Ed policy." Which was a nice way of putting it. I just wanted to let girls know it's best to wait until marriage, and that life is too valuable to take a gamble on.

As the Candie's teen spokesperson, I appeared on television and in magazines to talk about waiting to have sex until marriage. I simply wanted to tell American teenagers that saving sex for marriage was the only way to be 100 percent certain that you won't get pregnant. That's all. In the meantime, when I'd appear on television to talk about these issues, the media would invite Levi to comment on my statements.

He also appeared on *Larry King Live* (and other shows), along with his sister and mom. It was such a weird interview, because it's so hard to get Levi to talk about anything. He was inarticulate to the point of being painful, as Larry King struggled to fill in the conversational blanks.

When Larry King asked Levi if he had an attorney, he said no. Then, after the commercial break, he said, upon further reflection, he did. It was an entire interview of one-word answers and half-truths. Then Levi said, "Bristol doesn't trust my sister."

Though Sadie would give a ridiculous excuse ("Bristol and I don't see eye to eye, mostly because I have friends that Levi has dated"), Levi was right. I didn't trust her. And I no longer trusted a guy who used to be my snowmachining buddy but now had a spray tan and bedazzled skinny jeans.

Many completely false accusations came from his "media tour." He said that Mom called Trig "retarded," that she knew we were having sex, that she didn't know how to shoot a gun, that she put her career above family, and that he and I had sex in the Governor's Mansion!

After all of the lies he'd told me over the course of our relationship and the lies concocted about Mom on the campaign trail, I thought I'd lost the ability to be shocked. But I was completely floored.

The most easily debunked myth was the accusation that I wouldn't let him see the baby. I would've welcomed his help, and I'd asked him to help. I wanted to at least fix this part of the lies.

Let's get together and make a relationship happen between you and Tripp

I texted in May. It had been months since Levi had seen him, even though I repeatedly asked him to see Tripp. (On the advice of my attorneys, I kept records of all these attempts.)

When I didn't hear back from him, I went to my lawyers and we drafted a very generous agreement. I offered to let Levi have Tripp every Wednesday and every other weekend. Since this was an enormous amount more than he was currently seeing him, I thought he'd be pleased.

But we never heard back about the offer. After a week or so had passed, I texted him to find out if he'd even seen it. He texted:

Yeah, but I'm not agreeing to it, I want him 50% of the time.

Okay

I texted back, a little surprised.

Please come get him now because I have class.

Of course, he was busy *at the moment* . . . and every other time I asked.

On May 14, I got my chance to move slightly away from all of that custody drama by finally walking with my class in our Wasilla High School's commencement ceremony.

I had always dreamed of this graduation day. I'd seen the fun celebrations so many times in movies, the inspirational speeches, the way the principal moves the tassel from one side of the mortarboard to the other after handing you the diploma, tossing the hat in the air in unison. . . . It seemed so meaningful, so poignant, so wonderfully sad. I imagined my lifelong friends promising to go off somewhere together—and vowing to all be roommates. . . .

Okay, so maybe I'd built it up too much in my mind. After all, it had been a struggle. Going to school and taking online classes wasn't easy when I had a little, precious, screaming baby on my hip the whole time.

Our graduation was held at the Wasilla Sports Complex, and I was fine with *People* magazine covering my graduation. I wanted this accomplishment to define me more than my mistakes.

At the rehearsal the day before, I didn't really talk to anyone. I suddenly realized that my friends had moved on without me. They had boyfriends, I had a baby. And I was too embarrassed to make conversation with my guy friends. They didn't care that Tripp had puked on me before I walked out the door. To make matters worse, we were practicing our commencement march when I noticed that the girl in front of me with her lip pierced was one of Levi's girlfriends. (He'd even sold a story about how they were dating.)

On the night of graduation, we loaded up a bus at Wasilla High and drove to the Sports Complex. The cameraman from *People* somehow managed to sneak on the bus with me, and I laughed.

At the Sports Complex, as we were about to do the march, I approached Sammy. We'd been looking forward to walking together for years.

"I decided to walk with someone else," she said.

It devastated me, and I went into perseverance mode and just focused on getting through this night.

We marched in by walking around the track at the top of the facility. It was a huge event, and the place was packed. But all of the joy slowly drained out of the experience.

Since everyone wore the same red gowns and hats, you couldn't pick me out of the crowd. But as I watched all of the photos go by on the slideshow—photos of my friends goofing off and having fun—I felt like a total stranger who'd just come off the street to force herself into the festivities.

When I went onstage to get my diploma, I figured I'd look at my principals and have warm memories of coming of age under their wisdom and guidance. Nothing could've been further from the truth.

All of the drama that surrounded life in Wasilla really soured the way I felt about them. First of all, they had kept my former uncle Mike on the coaching staff after he called me—and other students—profanities. Second, they'd given quotes to magazines about me, which seemed like a particularly unusual and unethical thing for principals to do . . . especially since they mysteriously told one publication that I'd been absent from school and transferred to a different school. Third, they let reporters come to our school and offer money to the students on school property to talk about me.

"Hey, do you have any dirt on Bristol Palin?" one British tabloid reporter asked a friend. When she said that she didn't, he replied, "Want to make something up? I'll pay you."

But graduation was supposed to stop all of this nonsense. Mom was the speaker at Track's ceremony, and I was looking forward to hearing her speak at mine. But two weeks before our big night, they canceled their invitation. The principal said he didn't want the night to be about politics.

As it came time for me to get my diploma, I looked out into the audience to find my family. They were sitting in the first two rows, but I didn't see Tripp. I was worried that he'd scream, that he'd need his diaper changed, that he'd make a commotion when he saw me on stage.

"Where's my baby?" I whispered to Mom. She pointed back to the door, where Aunt Heather was holding him and feeding him a bottle. When I saw that precious bundle in my aunt's arms, I realized that he meant so much more to me than this ceremony or piece of paper. Still, that diploma was the first step toward a better life together. I waved at him before striding across the stage to get my diploma. I had a 3.497 grade point average, which kept me just shy of being an honor student.

Not bad for an unwed teen mom.

The next week, I was on the cover of *People* magazine, with my baby on my hip and a diploma in my hand. I was the only teen mom to graduate with my class.

Chapter Eleven

Already Ben There

Even though I was a proud high school graduate, I walked around town with a scarlet letter on my chest. At least that's how I felt. I'd read Nathaniel Hawthorne's novel in tenth grade but had only recently developed a newfound appreciation of poor Hester Prynne's struggle to create a new life for herself after getting pregnant outside of marriage.

After all, who could ever want me? I felt damaged, tied down, and unworthy of anyone's love.

But I remained hopeful, and when my friend Ben came over to borrow some snow pants of Track's, I felt like maybe I could find love.

I happened to be home, so I answered the door. When I saw him, it was like no time had passed, even though it had been a year since I'd run to him with a broken heart caused by Levi.

Our chat turned into hanging out, and it felt so good to be seeing him again. He even broke up with his girlfriend, which I took as a sign that he was really interested in me. Ben and I went to Kenai with my family on a fishing trip. We'd always scour Craigslist for houses. Not necessarily to live in, but I always wanted to find a good deal for investment purposes. Though I thought things were progressing nicely, I realized I was wrong. He went to the prom with one of my former best friends, Chelsea.

There was another guy, Gino. When we started talking, he broke up with *his* long-term girlfriend, too. He took me to Whittier on a family fishing trip and I so loved his family. That's when I thought, *Why would God bless me with someone this good?* Again, I took the fact that he broke up with his girlfriend as a sign that he was really interested in me, too. (Oops! Wrong about that as well. After we started dating, he actually was with his ex-girlfriend behind my back.)

Though nothing was serious with either of these guys, I enjoyed hanging out with them and loved the feeling of being "single" after so many years. Well, I loved the idea of it more than the feeling. When I found out both of these guys had been hanging out with other girls and even Levi at the time, it devastated me. I felt like there was no place to run, to get out of Wasilla's tangled web. It was the most fragile I've ever felt. With very little sleep, a demanding infant, and a lot of shame, I felt so down in the dumps.

But like most teen moms in America, I didn't have time to wallow in my emotions. I had to put one foot in front of the other and figure out my life. Immediately, I enrolled for the fall semester in nearby Mat-Su College for three hours a day and took a public speaking class, which just happened to be a required course. That semester, the instructor asked us to do a presentation, so my

PowerPoint was called "How to Properly Swaddle a Baby" . . . a valuable lesson Dad had taught me after Tripp was born. My part-time college schedule allowed me to go to school and work two or three days a week at a local coffee stand. I was starting to slowly build a future for myself, but I longed for a place of my own. It's not like my home was unpleasant or uncomfortable. I'd outgrown my childhood room now that it was stuffed with a crib and baby toys. Because I was a mom, getting older, and didn't want to rely on my parents, I wanted to try living on my own.

"Hey, Bristol," Mom said one morning as we ate breakfast. "We're going to build a shop for Dad."

He'd always wanted a place to store his plane and our snowma-chines, and our family—with the addition of Trig and Tripp—was quickly outgrowing our house.

"Cool!" I said. "Build me an apartment, too!"

Amazingly, Mom didn't blink an eye. She and Dad thought it would be a great idea to add a living space next to the shop. That would give Tripp and me a place to live, and it would help me take care of Trig, Piper, and Willow, since I was the only sibling who had her driver's license.

They let me design the layout for my 1,500-square-foot apart-ment.

"I want the bed here, the kitchen here," I remember saying happily. I planned two bedrooms right next to each other, a deep sink so I could bathe Tripp in it while he was an infant, and a big bathtub for his baths when he got older. It would be such a blessing to have a place to stay and I was thankful!

And then, out of the blue, Levi texted me.

Hey, can I have Tripp?

It sounded suspicious. He never asked to see his son unless he was up to something. In May, he had asked to see Tripp, only to have me find out a few weeks later that *GQ* had been in town for a photo shoot. Levi had posed, shirtless of course, while changing Tripp's diaper. The camera took the photo when Tripp was completely naked. Not just cute "baby bottom" naked, but complete frontal nudity of my innocent baby son. I was furious!

So when Levi wanted to see Tripp, I immediately texted back,

What magazine's in town?

Sure enough, he had planned on using our son as a prop in yet another photo shoot. It was almost too much to bear. Levi was constantly drumming up drama, pretending to be a good father for the cameras and leaving me with the diaper bill. However, I couldn't spend too much time dwelling on his shortcomings. Instead, I simply hugged Tripp a bit closer at night and tried to focus on being the best mom I could be.

In the middle of June, Mom called with great news that really lifted my spirits. "Bristol, what would you think if I was done with being governor?"

I didn't have to think twice. When Mom went back to Juneau after the campaign, she'd been absolutely bombarded with false lawsuits, ethics violations, and accusations. Her administration was hit with hundreds of "fishing expeditions" in the form of Freedom of Information Act requests for months of e-mails among Mom, Dad, and her staff. The people requesting the information weren't concerned citizens worried about transparency in their government. Instead they were Mom's political enemies looking for something—anything—that might make Mom look bad. One of these requests created twenty-four thousand individual sheets of paper, due to the necessary printing of the e-mails, copying them for the attorneys and governor's staff to remove confidential information from them before organizing and reassembling them. That's just one of the FOIA demands. There were hundreds, and they kept coming.

Suddenly, Mom and her staff were buried in expensive paperwork . . . to the tune of $2 million of taxpayer money. Now, if Mom would "unbudget" a chef that could make gourmet grilled cheese in a single bound, she wasn't going to stand by for this kind of waste. Even though every ethics charge filed against Mom was either tossed out or found no wrongdoing, a cloud hung over her head.

Normal citizens didn't understand these dynamics. Why would Mom have such high legal bills if she was innocent?

The key was this: anyone anywhere could file an ethics complaint for absolutely no money, illegally leak the complaint to the press, and then sit back and watch the destruction—at no personal cost. However, each complaint had to be formally processed, and Mom had to pay for her own legal defense. (It cost Mom and Dad $500,000.) Her approval rating dropped from nearly 90 percent before she went on the campaign trail to 56 percent.

So when she asked me what I thought about her stepping aside and letting the lieutenant governor take over that last lame duck year of her term, I was all for it. As far as I was concerned, we'd lived through that chapter of our lives and it was time to move on. In fact, everyone in my family and Mom's true friends agreed that stepping aside was the right decision for the state.

It was a sunny Fourth of July weekend when she held a press conference in our backyard. There was one microphone on a make-shift podium, a bunch of press, and an audience of friends and family wearing short-sleeved shirts. Dad stood beside her wearing jeans, while Piper held Trig. Baby noises could be picked up through the microphone that would broadcast this speech—one of her last as governor—to millions. She had eighteen months left in her term, and they all came to an end in an eighteen-minute speech.

"I polled the most important people in my life, my kids, where the count was unanimous," she said. "Well, in response to asking, 'Hey, you want me to make a positive difference and fight for all our children's future from outside the governor's office?' it was four yeses and one 'Hell, yeah!' And the 'Hell, yeah' sealed it."

That "Hell, yeah" was probably me.

The nation, once again, was shocked by my mother.

That summer I was working at my job at the coffee shop with my cousin Lauden. Because I made only $7.25 an hour, it was hard to also pay a babysitter. So that day—as she did when class wrapped up and I worked in Anchorage—Aunt Heather kindly agreed to watch Tripp for me, before she went to her job. Lauden and I had poured many skinny mocha lattes that day, when a woman came and placed her order. On her way out, she said, "Hey! We're looking for a receptionist at my father's dermatology clinic and spa. If you're looking for a different job, we have an opening."

"No way!" I said, hardly believing my ears. "That would be so much fun." Her father was Michael Cusack, a successful dermatologist who'd been practicing in Anchorage for decades. He had even owned the Alaska Aces hockey team in Anchorage at one time.

"Bring your résumé and we'll have an interview," she said as she took her espresso and put it in her cup holder. "By the way, we pay $15 per hour."

And with that "by the way," my heart soared. With that kind of salary, I could afford to pay a babysitter and have a regular schedule. I went into the Alaska Dermatology and Laser Center at nine o'clock in the morning and shook the hand of Dr. Cusack, who was such a fixture in the medical community of Anchorage that his office had more than ninety thousand charts.

He knew who I was. You can't live in Alaska—or anywhere now—with the last name of Palin without people recognizing you. He told me what he was looking for and asked me what I wanted to do. "Listen," he said kindly. "I've seen what you've been through, and I want to help you create a stable income for yourself. If you do good work, you can move up in seniority here. When can you start?"

"Tomorrow?" I responded. My public speaking class had just wrapped up, and everyone in the class now knew more about swaddling babies than they ever wanted to know. Plus, I had also developed speeches about the benefits of waiting for marriage to have sex, about Down syndrome, and about the steps for decorating a house! That allowed me to start at the clinic immediately, and I was on the road to a much more stable life.

The ladies who worked there became my instant friends. Marina was a medical assistant in her mid-twenties, Crystal was the office manager in her early twenties, and Janice, in her forties, darted around doing all kinds of things in the office. I was told I'd also do whatever was needed at the time. Though all of us were so different, we shared a serious commitment to having fun.

All kinds of people came into the office—for skin cancer issues, rashes, rosacea, Botox, skin peels, and microdermabrasion. We had exam rooms, and I'd bounce around to each room helping everyone. After people signed in, I'd frequently be the one to take patients back to the exam room to ask some basic questions, like their name, their medical history, and why they were coming in that day. After I'd leave the room, Marina would come in to check their ailment.

Sometimes, after I was safely out of earshot, they'd ask her, "Is that Bristol Palin?"

Marina would lean in conspiratorially and say, "She gets that a lot."

I could deal with a lot of issues—cysts, stitch removal, rosacea, eczema. But the first time I learned about Botox injections, it was a little too much for me to handle.

"So when we hit the bone, we hear a sound like this," the doctor was explaining to me as he showed me the process. When I heard the needle go in and make that sound, I got a little light-headed.

"Oh no," said Marina. "She's having a vasovagal reaction."

I didn't know what that was, but it didn't sound—or feel—good. They took me by the arm, and I had to go lie down in Dr. Cusack's office with my feet elevated. It wasn't the last time I'd be embarrassed at the office. One night, I dyed Willow's hair but didn't wear gloves. The next day, my hands were absolutely purple, like plums. And nothing would get that stain off—bleach, peroxide, alcohol wipes, nothing. So for two weeks, I had to endure the jokes of my coworkers as I worked in the dermatologist office with hands as purple as Grimace from McDonald's.

The questions became even more hushed. "Is that Bristol Palin with some strange skin disease?"

People didn't realize that the daughter of Todd and Sarah Palin would work at an eight-to-five job, doing normal things . . . as if the RNC would suddenly pay for my electric bill and ever-growing diaper expense! So we had a lot of fun with my easily recogniz-able face. Dr. Cusack would call me by a different name to throw people off. He'd say, "Susie, please go get those charts."

Since Dr. Cusack was very set in his ways, we'd pull pranks on him. He always wore blue surgical masks while doing procedures. Once, Janice drew a smiley face on the outside of it, so he put on his mask and had a big Sharpie-induced grin on his face without knowing it.

He also had a favorite Chinese restaurant that made a waffle house look like a gourmet restaurant. I couldn't believe that some-one as sophisticated as the doctor would purposely choose to eat there.

"Hey, want to go to lunch?" he'd ask us. "Just follow me, I know a really good place."

Every single time, no matter how much we complained, we'd end up at that old dusty restaurant, where the waiters knew exactly what he'd order and we'd sit on the glass-covered porch and try to grin and bear it.

My mom is always preaching "Man was created to work!" And that certainly rang true for me throughout this entire time. For the first time in my life, I felt stable because of going to work every day. There's something about having a job and a sched-ule that made me proud to be setting a good example for my son and affording his clothes and toys.

Yet there was one loose end, and after so many months of trying to make a steady visiting schedule for Levi and Tripp, I decided to file for sole custody of my son. After all, I was the only parent with a job.

I'd grown so tired of my personal dramas being discussed on television and newsstands that I hoped to figure out the custody issue in the privacy of my own misery. My lawyer and I asked the judge to allow us to file using fake names, in order to keep the records from being made public and picked up in the news. Initially, the judge issued temporary orders, which limited access to our file and allowed us to file under John and Jane Doe. I was so relieved!

But then Levi—and his attorney, whom he met while representing his mother in her drug-dealing charges—went public with the case. He said,

> *I do not feel protected against Sarah Palin in a closed proceeding. I hope that if it is open she will stay out of it. . . . I think a public case might go a long way in reducing Sarah Palin's instinct to attack and allow the real parties in this litigation, Bristol and I, to work things out a lot more peacefully than we could if there is any more meddling from Sarah Palin.*

When a reporter asked what he meant by "meddling," Levi couldn't produce one example. That's because Levi wasn't afraid of my mother. He just knew that a long, drawn-out custody battle would damage our family's public image, create more drama, and provide opportunities for Levi to make more money. Plus, he could use the publicity to sell more articles.

And our problems were once again thrust into the public spotlight. On top of all this, he still almost never saw Tripp. As Christmas approached, I started feeling bad about the fact that my son couldn't even recognize his own father. That's when I texted Levi and made a suggestion:

Let's take Tripp out together for his birthday.

I knew I shouldn't have, but it just felt weird that a big mile-stone—like a birthday—would come and go without a dad there. I tried to keep it on the down low. If Mom and Dad—or anyone—found out I'd reached out to him, they wouldn't have understood. When Levi agreed, I chose an Anchorage Red Robin as a place to eat. We went on a weeknight when it wasn't busy, around nine o'clock, to avoid anyone who might know us.

This was the first time we'd been in the same vicinity in a long time, so I was interested in how he'd respond to seeing me and his son. I'd wanted to work out a time for them to see each other, but Levi just never seemed like he cared. The fact that he was willing to meet us seemed like a good development.

Would it be possible to actually have a family? To get married and have a life together? Could he possibly be sorry for spreading lies about my family on national television? Had he changed?

He slid into his seat across from me and Tripp, with hickeys all over this neck.

"How disrespectful!" I said to him in disbelief. "Why would you show up for your son's birthday like that? At least cover those up with a scarf or something."

"I got them from some drunk girl," he said, as if that made sense. As if the alcohol on the lips of an intoxicated girl is somehow magnetically attracted to his neck. It was the same situation as seeing Levi three years ago with hickeys.

I somehow managed to get through the meal in spite of my disappointment. But the first person I called when I got home was Ben.

"Would you believe he'd show up looking like that?" I asked.

Ben paused before answering. "Well, you know," he said, "Levi *has* been dating a girl with a kid for a while now."

My heart stopped.

"A kid?" I asked. "He's been dating someone with a . . . kid? How old?"

"The same age as Tripp."

I couldn't believe my ears. Why would Levi hang out with someone else's baby and not with us?

I broke down completely.

How could he do this to me? Why did I think we could make it work? Why did I take Tripp out past his bedtime just to meet up with such a tool?

I trudged to work the next day, and the next, as I tried to forge my way with Tripp in this world. I was thankful for my work as "Susie," and enjoyed my relative anonymity in Anchorage.

Because I now had a regular job with a nicer salary, I decided that I wanted to move closer to Anchorage and have my own place. The commute from Wasilla to Anchorage wasn't too bad. On a good day, it was less than an hour, but snow and ice could sometimes make it more than two hours. As you know, Alaska sometimes has a little snow and ice. Because I'd earned a little bit of money through endorsement deals, didn't have to pay Mom and Dad rent, and had always been cautious about my budget, I had saved enough money to buy a condo closer to work. So I approached my parents about it.

"Really?" Mom said, her nose scrunched up the way it does when she's not happy. "But what about that apartment for you and Tripp? You haven't even finished picking out the colors for the wall."

She was right, of course. They'd already begun building this great place for me, though it would also house Dad's plane and later Mom's studio. I was a big part of how the family worked since Willow didn't have her driver's license yet and Track was stationed in Fairbanks. I ran errands and did the grocery shopping in exchange for a free roof over my head. It was a generous arrangement that had allowed me to get my high school degree.

Even though I knew it would disappoint my parents—and complicate their lives—I was sick of the drama of Wasilla, sick of the daily commute, and ready to live on my own. Ben and I were still friends, even though he took my friend to her prom. I decided that we could still hang out as buddies even though romance was not in our future. He was a lot of fun to spend time with, especially because he was game when I suggested calling in sick to work one morning, dropping off Tripp at the sitter's, and shopping at the mall. It felt so care-free to be shopping without carrying the baby and a diaper bag. Then we drove by condos that we'd seen on Craigslist. It was so much fun to drive by each one slowly, trying to imagine if each could potentially be my first home! Coincidentally, Gino—who followed in his father's footsteps as a realtor—had e-mailed me the MLS post on a foreclosure that had only been on the market for one day.

In my real estate excitement, I texted back:

Come open it up for me so I can see the inside!

He drove into Anchorage from Wasilla and was surprised to see me sitting in Ben's truck. That's the thing about small-town life; everyone is intertwined. Suddenly, my old flames were face-to-face, all so I could peek inside a condo.

It had three bedrooms, two and a half baths, nineteen hundred square feet, with a two-car garage. It had cool high ceilings and wasn't a cookie-cutter apartment. It had character. By the end of that day, I had already made an offer on the place. When the bank accepted my offer, I was so proud to be able to own my first home at age nineteen—through hard work and careful management of the money I earned from work and my endorsement deals. Because I wanted to protect my "new beginning," I went to great lengths to purchase the condo privately. I conducted the sale through my attorney's law firm under the name SM Properties. That way, I figured, I could get out from under the shadow of all the politics and family connections, to start in a new place with my new job.

I was thrilled!

Because my parents weren't enthusiastic about me moving, I couldn't exactly ask them for help in moving my things to Anchorage. However, Ben showed up, borrowed Dad's trailer, and single-handedly moved all of my furniture—my mattress, my bed, my couch, my dishes—from Wasilla to Anchorage. He didn't stop there. He moved all of my things into my new condo, which was three stories. Now, that's a good friend!

I immediately started making the condo uniquely mine—with large leather couches, flat-screen televisions, neat rugs, pink kitchen appliances, and—of course—a toddler bed for Tripp. I painted the walls purple, bought a leopard print carpet for the upstairs hallway, and a purple-and-black swirled carpet for the living room. In other words, I made it into a comfortable—yet hip—place for a nineteen-year-old to develop into "home." I fully expected to have to replace Tripp's tiny bed with a twin, and then with a full bed as he grew up right there in Anchorage.

But I soon found out that even living there was not enough buffer between me and all of the Levi-produced drama.

"Puke!" I said, when I found out my son's father had posed nude for *Playgirl*. I couldn't imagine that the stories that were circulating about my life on the Internet now suddenly had an element of porn to them.

I was determined to move on with my life, and I put Levi out of my mind as best I could. When in the spring of 2010, *Harper's Bazaar* contacted me to do a magazine shoot, I jumped at the chance. It seemed like a fun way to show the world that I'd moved on from the tabloid messiness that had defined my life. They arrived with an entire camera crew, wardrobe, makeup, and even an array of cakes, cookies, strawberries, and other beautiful confections made in Los Angeles and brought to Alaska just for the shoot. The gowns they brought were jaw-dropping. I felt like a princess as I slipped off my jeans and slipped on a Lanvin gown that cost more than $4,000, a Carolina Herrera shrug that cost almost $7,000, and an Isaac Mizrahi gown that cost almost $15,000. It was hilarious to be standing in my kitchen while my hair was curled and sprayed into place, all while Tripp was smearing cake all over his chubby cheeks.

The reporter was there at my condo while we celebrated my brother Trig's birthday. I'd decorated the condo for the event, with hand-lettered signs and balloons. Mom showed up and chatted with the reporter. It was a fun experience, one that focused on my new self, my new life, and had very little to do with that guy named Levi.

During this time, Ben was a constant friend. Well, he became more than a friend as he listened to all of my complaints and absorbed a lot of my heartache. And we officially started dating after I ran into him around Tripp's first birthday. Ben is a soft-spoken, hardworking guy. I actually used to babysit his little brother and sister, so we'd been friends for a long time. But the main thing that attracted me to Ben was that he seemed to really love Tripp. I was astonished how easily he played with my son. He'd change his diapers and help me put him down to sleep. We made Costco runs together, he helped me assemble my computer desk, and he stopped by in the evening on the way home from his job near my house. Later, when Tripp was a little older, Ben would put him on a snowmachine and we'd laugh as he tried to ride by himself. He'd stop and go, stop and go. But the whole time, Ben was right there beside him making sure he didn't fall. It was refreshing to have someone help bear the weight of the responsibility of having a baby. We were inseparable.

Well, not completely inseparable. One day we had a little fight—just a silly argument, but we decided to take a short break to sort things out. We still stayed in touch and we knew we'd be back together as soon as some time had passed.

That's when my attorneys told me I had to get serious about getting Levi's visitation schedule set up. Levi kept publicly saying that I was keeping him from seeing his son, and my legal fees were stacking up.

"Okay," I said to my lawyers. "I'll take care of it."

After all, I wanted to put the Levi part of my life behind me as much as anyone.

Chapter Twelve

Home

Come to Anchorage to set up a schedule to see Tripp.

In spite of the fact that I never really wanted to see Levi's face again, I texted him when my attorneys suggested in May 2010 we figure out—once and for all— our custody arrangement. No progress had been made in a year, lies were still being told, and Tripp was still without a father.

I'm not doing this forever, Levi.

A man of very few words, he immediately texted me back.

Right.

Hardly a surprise. He was always promising things on which he never followed through. But I *was* surprised when he actually showed up the next day in Anchorage.

I opened the door with Tripp on my hip, expecting Ricky Hollywood, with his spray tan and skinny jeans and a European handbag full of excuses. But instead what I saw was an unshaven Levi wearing boots, his camo jacket, and a camo ball cap. He looked very much like the guy whose locker was beside mine in seventh grade.

"Come in," I said, motioning inside my new condo, thrilled to be able to show it off. I'd been able to care for Tripp in a nice comfortable setting in spite of Levi's lack of child support and visitation. Track's girlfriend, Britta, lived with me, because I felt a little scared about living by myself. Nonetheless, I felt like I'd made it "on my own." Levi seemed impressed by the tour of the place, and he ruffled Tripp's hair.

It was a little awkward to have him there, after all of the bitterness that had passed between us like a bad case of the flu. "Hey, let's take Tripp for a walk," I suggested. At least that would entertain the baby while we hashed out custody issues.

"So what happened to Ricky Hollywood?" I asked. We were walking on the sidewalk pushing Tripp in his stroller.

"I gave that up," he said. "It wasn't really me."

"And what was up with your hair, anyway?"

He took my teasing good-naturedly, and we laughed as I made fun of his ridiculously fake persona. It was strangely relaxing and comfortable. Tripp laughed and played quietly, while I asked Levi about his late child-support payments. The word *late* implies that he would one day pay them, which we both knew was not the case. Though he'd made a pretty good amount of money selling his stories to the press and his body to *Playgirl,* he had wasted it on toys, fishing gear, and a hefty percentage to Rex and Tank. By now, he owed me more than $20,000 in child care, and I suggested he just sign over his rights and call it a day.

But the heavy conversation soon yielded to the fun in the park. Suddenly, instead of a harsh interaction between enemies, it felt more like we were just two parents passing a nice day in the sun. If you didn't know us and were just a stranger walking by, you might have mistaken us for one happy young family. When we said good-bye, I didn't hug or kiss him. However, I felt just briefly the sensation of the way things used to be . . . before life got so darn complicated.

That night, after I'd put Tripp to bed, my phone buzzed with a text message.

I miss you, Bristol.

Initially, I was shocked. Here was a guy who'd so publicly betrayed my family we probably shouldn't have even been in the same room. Someone who'd sell his body to make a buck. Someone known primarily for lies and mistakes and sleaze.

And yet . . .

I sat there alone, and I wondered if I'd really given "us" a chance as a family. Had I really considered Tripp's best interests when I'd tossed Levi out on his butt so many times before? Now that I had a child, didn't Tripp deserve to have his father in his life? Could Levi have really yanked himself from the old persona and seen the light?

It's hard to explain what I did next. Women throughout history have looked back at their relationships and wondered how they could've loved a man who treated them so badly. From Hillary

Clinton to Sandra Bullock, from Jackie Kennedy to Jennifer Aniston, from Elin Woods to Princess Diana. My story is—sadly—not unique. And neither was my response. I texted back:

I miss you too.

But this time, I told myself, I'd be in control of our relationship. This time, instead of reeling from his infidelities, I'd stop them before they happened. This time, instead of watching him indulge in his teenage obsessions like fishing and hunting, I'd help him learn how to grow into a man and an attentive father. Levi's dad hadn't been the best role model: he'd had a long-term affair when he was with Levi's mom, and eventually left her for the other woman. Perhaps this was the reason Levi had abandoned me and Tripp. He simply never learned how to act in a mature relationship.

That was about to change.

The next day, he came over—Levi again, not Ricky—and we sat together commemorating our new relationship. But instead of celebrating with a candlelit dinner, we began it with a spiral notebook and a ballpoint pen. I was making a list.

"Okay, if this is going to work out, we have to agree on some terms," I said. Normally, I'm not a bossy, controlling girlfriend, but if I was going to risk upsetting my family for this guy, it absolutely had to work. One way to guarantee that everyone was on the same page was to spell out his obligations and my expectations. Here is my list, almost verbatim . . . (I edited it to eliminate the profanity, which I'd added to make a point to Levi!)

1. Protection: I shouldn't feel threatened by your family. You should protect Tripp and me before anyone else.

2. Respect: I don't deserve to be treated poorly and called a b—ch.

3. Forever love: Not just now, but forever.

4. Willingness to change: I can't hold your hand and guide you through life.

5. Love for our child: You have to try to be a good dad to Tripp.

6. Appreciation: Be grateful for the two of us.

7. Dedication: Try to improve yourself for our family.

8. Honesty: Commit to me and only me.

9. Devotion: Be here for us consistently.

10. Stability: When are you getting a job? An education?

11. Motivation: Start doing stuff for yourself.

12. Maturity: Spend time with Tripp, not just fishing 24/7. You can't spend all your money on fishing gear. Have to help out with the bills.

13. Controlling your anger: You know you have a temper.

14. Regret: I want you to feel regret for everything you have done to me. Regret for how much you have disrespected my entire family. Regret for missing so much of Tripp's life.

15. Apology: You need to apologize to my family, publicly and privately.

16. Education: You need to get your GED. I tried to push you into doing this, but you have to have motivation yourself.

At the bottom of the paper, I wrote, "I've given up all of my life to be the best mom I can be. Why do you go fishing without even thinking twice about it? Why do I feel guilty about spending $200 when you throw money down on pointless gear?"

But the part that haunts me now is the last sentence, on a totally different page. "Why don't you have emotion toward Tripp and me?"

That question should've caused me to pause, to protect my son, me, and the rest of my family from Levi. But writing it down on paper made me feel like I owned the situation just a tad more and gave me a false security of "setting the terms." This time, I thought, things would be different. This will be the moment we look back on when we're old and gray and sitting in our rocking chairs, and we'll laugh and say, "Yes, that's when everything turned around for us . . . after all the chaos of the campaign, after all the immaturity of high school, after our love for Tripp transformed our love for each other . . . it all came together after the day we took our son for a walk."

He quickly agreed to all of my stipulations. He promised to get a job, get his GED, and quit having that weird relationship with his obsessed sister, who was always trying to hook him up with her friends and had physically threatened me when I was pregnant. Most of all, he said he would apologize to Mom and Dad—publicly and privately—for all of his lies.

Suddenly, life seemed a lot "fuller" as our almost-but-not-quite family began trying to make things right.

Some time after Levi and I got back together, I went to the mailbox and found a card, addressed to "SM Properties, LLC," in a yellow envelope. Inside was a white piece of copy machine paper, with this message written in pink highlighter:

B.P. –
We know . . .
And so do they . . .

My heart sank. Who would know that "B.P." was really SM Properties? And what does "so do they" mean?

I was about to find out.

Suddenly, I started noticing strange people hanging out in front of my home.

One day I was home giving Tripp a bath when someone knocked on my door.

"We can do this the easy way or the hard way," the man said. He looked like a homeless man. "The hard way is that you avoid me and I still get the pictures. The easy way? You cooperate and I'll give you a cut of my profits."

I couldn't believe my ears. He was asking me to pimp myself out.

"Let's do it the hard way," I said before slamming the door. After he left, I leaned up against the door and realized I really couldn't hide.

The hard way suddenly became even harder. The local AP press started showing up on my doorstep every day. Larry King's producers sent me packages—suspenders for Tripp!—to entice me to come on his show. Some despicable bloggers drove by my house every day and would keep a log on what cars or trucks were in the driveway. And speaking of blogs, another one popped up. This time, it was Levi's sister, who created it to falsely answer any question that people around America had about me.

Looking back, I realize it couldn't have been a coincidence that everyone found out where I lived right around the time I got back together with Levi. But I didn't put two and two together at the time. Then, I was simply dedicated to the idea that finally Levi, Tripp, and I would become a family.

And he seemed to be, too.

I was at work when Levi went to Mom and Dad's house in June to apologize. I gave Levi my mother's phone number and he texted her to arrange a time to stop by. All of the lies he'd said about them needed to be addressed both publicly and privately, so he was taking care of the private apology first.

Mom and Dad allowed him into their home, and they sat at their kitchen table. Willow, as always, was eavesdropping from the balcony and kept me apprised via text of all that was going on downstairs.

"I'm sorry about all the lies I've made up about you," he said. Mom seemed to take it better than Dad . . . at least according to Willow's play-by-play. Dad was not impressed by the fact that Levi was dressed like his old self and claimed not to be under the influence of Rex and Tank anymore. And, according to Willow, Dad was not going to easily let Levi off the hook.

"I'm the one who's been getting up with Tripp in the mornings, taking care of your son. I'm the one who's been comforting Bristol. I'm the one who's been changing the dirty diapers and watching your son take his first steps. Where have you been?"

Levi apologized, again, and said that he'd have to prove himself to gain their trust. It didn't go flawlessly, but at least he had shown that he could own up to his mistakes and he promised to make a public apology. My parents were skeptical.

But, within days, *People* magazine ran the note that Levi and I had drafted that day on the notebook paper:

> *Last year, after Bristol and I broke up, I was unhappy and a little angry. Unfortunately, against my better judgment, I publicly said things about the Palins that were not completely true. I have already privately apologized to Todd and Sarah. Since my statements were public, I owe it to the Palins to publicly apologize. So to the Palin family in general and to Sarah Palin in particular, please accept my regrets and forgive my youthful indiscretion. I hope one day to restore your trust.*

It was a Band-Aid on a gaping wound of problems he'd caused my parents. While I took it as proof he'd really put his old ways behind him, they couldn't come around to trusting him again.

When shortly after that I drove over to the house to see my folks, ready to have a frank and honest conversation with my parents about my decision to get back together with Levi, there was tension in the air. I soon realized a calm conversation would not be happening that day.

When I pulled into the driveway, Dad came out and flagged down the truck.

"Why do you keep going back to a guy who lied about us so badly?" he said. "Do you know how many times you've come crying to me, upset about that guy? He's done nothing but disrespect this family."

"I'm trying to give Tripp a father," I protested, but I could tell by the look on his face that he wasn't ready for me to explain why I'd chosen to reunite. I didn't get a chance to explain that I wanted Tripp to have a family as good as the one I had growing up.

My heart hurt, as I processed the fact that the damage I'd done to my family was permanent. By this time, I was crying.

"He has betrayed you over and over and over. It will happen again," he continued. "I'm done talking to you."

I drove off, rationalizing my decision with every mile. *I'm going to have my own family now,* I thought. I pressed on the gas to get to the only person on the planet who understood me. Levi was in the same boat as I was, because he'd forsaken his family to be with me. In a Romeo-and-Juliet-type way, we were pursuing love in spite of all the forces that were coming against us. What could be more romantic?

In July, Levi made the romance official when I came home from work and walked into my bedroom to find he'd made a heart out of rose petals on my bed. There were also bouquets of red roses sitting around.

Again, I don't even really like flowers—they require so much attention, and the best-case scenario is that they die slowly! The fact that he didn't know my antiflower policy probably suggested he really wasn't all that into me. By that time, I wasn't even physically attracted to him. After all of his betrayal, his face, his body, even his aroma just seemed "off" to me. But my instinct was still to move toward him in an effort to become a real family.

I didn't notice the little box in the middle of the heart. That's when he got down on one knee and said, "I want to be with you. I'm sorry for everything I've done. I want to be a family with you and Tripp. Will you marry me?"

"Yes," I said, and he slipped the ring on my finger.

I was fully aware of the shock waves that engagement was about to send across the nation. So we started making plans.

We took out our notebook and began writing a public apology, one that I faxed to my attorney as we prepared to figure out how to go public with our relationship. Instead of allowing it to filter through Wasilla, Anchorage, and then to the national press, we decided to take control of the story. Rex and Tank suggested that we sell an exclusive story to *Us Weekly*. Since people were going to find out anyway, why not tell the story in a way that not only benefited our new family but also controlled the message? They also suggested we fly to Las Vegas and get married, a suggestion I'm thankful I had enough sense to ignore.

Immediately, the magazine sent a contract over and we began working out the details. We agreed to a two-week exclusive, which meant we couldn't talk to any other media outlets about our upcoming nuptials. If we made the news or talked to other magazines before two weeks after the magazines hit the newsstand, we'd violate our contract.

And so we were quiet about our decision to get married, and I dreamed of what kind of wedding we might have. I also began sticking up for Levi in front of friends, my family, and in front of Dr. Cusack, telling everyone he'd really changed.

On July 14, I was sitting at work in Anchorage when I received an e-mail that would change my life even more for the next few months. It was from Deena Katz, a casting guru from Los Angeles who was finding people to be in the eleventh season of *Dancing with the Stars*.

"Dr. Cusack, can I go talk on the phone in your office really quick? I just need to make a call."

I got up from my desk, ducked into my boss's office, and called my attorney. I was thankful to finally be communicating with him about something other than custody issues!

"I can't believe this," I said, looking at the e-mail.

"Ask them to fax over the contract. Are there any negotiations?"

Dancing with the Stars is ABC's ratings juggernaut. In this twelve-week show, "celebrities" are paired with professional dancers who give a crash course in how to do dances like the waltz and the rumba. Let me be the first to say, however, I don't consider myself a celebrity at all. In fact, as I stood in our office with my phone pressed to my ear, my attorney asked me what I thought about participating.

"Well, I can't dance, and I'm not a star . . ."

But I didn't have time to really think about it. Soon, the magazine with the story about our engagement would hit the newsstands, and I'd have a lot of explaining to do!

The magazine with our announcement came out in the lower forty-eight about a week before it came to Alaska. But the magazines didn't have to be physically on the shelves for all hell to break loose.

The media had a field day with the newest chapter in the Levi/Bristol saga. Newscasters announced our engagement while playing the "Wedding March" or "Reunited, 'Cause It Feels So Good" as they discussed it. They took cameras to the streets, as normal people expressed disgust or surprise, or gave the occasional thumbs-up. Everyone everywhere was completely shocked.

Mom hastily released a public statement:

> *Bristol, at 19, is now a young adult. As parents, we obviously want what's best for our children, but Bristol is ultimately in charge of determining what is best for her and her beautiful son. Bristol believes in redemption and forgiveness to a degree most of us struggle to put into practice in our daily lives. We pray that, as a couple, Bristol and Levi's relationship matures into one that will allow Tripp to grow up graced with two loving parents in his life.*

But privately, she and Dad were furious. They were right in the middle of filming the TLC show *Sarah Palin's Alaska* when the news broke.

I called Track.

"Listen, don't come around here," he warned. "Everyone is so upset about this. Even Grandpa is very disappointed."

"Even Grandpa?" I was shocked. I was the firstborn girl in the family, which meant I automatically had a special bond with Grandpa and Grandma. He had always let me drive his four-wheeler, while I'd sit on his lap and he'd shift the gears. He'd also let me borrow his taxidermied tarantula and bat, and even a monkey skull, for show-and-tell every year. As a science teacher, he had accumulated a huge array of neat items like this! The fact that he would let me borrow them was a big deal because it showed how much confidence he had in me. That's why it hurt when Track followed up by explaining: "They say you can't be trusted."

That evening, after work, I met the only person in the world who understood. Levi stopped by the condo and I was looking forward to running into his arms for comfort.

But when I saw him, he didn't look happy to see me.

"I have to tell you something," he said.

His voice sounded more serious than I wanted on that day. Betraying my family had taken a severe emotional toll on me, and I didn't want to have to deal with another complication.

"Okay," I said reluctantly. "Go ahead."

"I think . . . I might've gotten someone pregnant."

I almost vomited. It was a pain worse than childbirth, worse than telling Mom and Dad about my pregnancy. But somehow, as I let the words sink into my mind, an eerie calm settled over me. My mind was spinning, but I had to figure out how deep the knife that had just been plunged into my heart might cut.

"Okay," I said. "Who?"

"Lanesia." This was the girl who chased me around the parking lot in eighth grade threatening to beat me up, the girl who always hated me, the girl his sister always pushed toward him.

"When did . . . *it* happen?"

"Back when she was house-sitting for . . ."

I remembered the house-sitting time frame and did the math. If he had been with her during that time, it meant she was pretty close to delivering! That couldn't be. There's no way Levi would've set me up announcing our engagement in a national magazine for such a humiliating fall.

"So that would mean she is due . . ."

"In two weeks."

There was no remorse, no comforting apology, no begging to let him stay around. He'd just taken the awful news and laid it at my feet like a heaping bag of trash I needed to take out. I was completely numb.

"Get out of my house. Don't call me, don't text me . . . I don't want to see you again."

When he walked out the door around seven o'clock that evening, I was—for the first real time in my life—utterly alone. Well, that wasn't quite true. I had Tripp. I looked at him, with his curly blond hair, his pudgy little hands, and his innocent blue eyes, and thought, *Well, it's just you and me now.* I put Tripp in the bed with me that night, and I watched him as he slept.

What have I done to us?

Dread filled my soul as I lay there in the dark, waiting for a sleep that wouldn't come.

Even worse, the "exclusivity" contract I'd signed with the magazine meant I couldn't tell anyone for another two weeks about the engagement plans . . . and that meant I couldn't even tell people it was off! Not that I had any shoulders to cry on. . . . I had betrayed my parents, my friends had turned on me. I'd just made a complete fool of myself and given my family the middle finger. Instead of talking to anyone, I wrote in my journal about it:

> *Why would I ever think he would change? I had this unrealistic fantasy that Levi wanted to be a family. . . . It made me turn on my own family, and let everyone who loves me down. I shouldn't have given in to that temptation. I pray for forgiveness for what I have just done and experienced, and pray for more wisdom and strength for my future.*

I went to work the next day in a fog of shame, yet I had to mask it.

"Congratulations," a coworker said. "We saw the magazine! When's the big day?"

"Um," I said, plastering on a fake smile. "We haven't quite set a date yet."

"Are you going to have a big wedding?" another asked. "We better be invited!"

This went on all day, with congratulations, questions about our plans, and excited smiles from patients who'd read the magazine and knew I wasn't really "Susie." Dr. Cusack's daughter, who lived in the lower forty-eight, sent me a huge bouquet of flowers.

I so wanted to have something legitimate to be excited about. Because of bad decisions, I didn't get to experience the wonder of my first sexual experience, the fun of telling an excited husband about a pregnancy, the expectation of a baby shower in a musty old church multipurpose room. The best I had was the hoopla over this phony engagement, every mention of which was a reminder that I'd betrayed my family.

It felt like I was stuck in a bear trap, with my bad decisions clamping down on me, keeping me in a painful condition with no way out. They were the longest and most terrible days. The only person who pursued me, the only person who wasn't either shocked, perplexed, or opposed to my "reengagement" was Levi. Despite my rejection of him, he texted me with updates, questions, and just conversation.

In one text he said:

By the way, I'm not sure I'm really the father of Lanesia's baby.

Again, daytime talk shows have nothing on my life. Lanesia denied it, too. But I didn't know what to believe. So I avoided him, ignored his texts, and didn't return his calls. All the while, I constantly had to answer questions about why Levi and I had gotten back together, about what color the bridesmaids' dresses were, and—of course—whether I was pregnant again.

During this time, when I had no one to talk to, my coworkers played such a key role in my life. They weren't offended by my decision and understood all of the complexities of a young single mom trying to figure out a path. Dr. Cusack, sensing I was emotional, offered me time off.

"Why don't you go home early to sort things out?" he said, kindly.

Oh, but how complicated the very notion of "home" had become. At one time, my home was in Wasilla, near Lake Lucille, tucked amid the birch trees and the moose that would sometimes block the driveway. Now? I had to go to a condo stalked by creepy photographers, a place that reeked of bad decisions.

I got in the truck, put my hand on the steering wheel, and turned on the ignition, just as my phone buzzed to alert me of a text from an old high school friend.

Lanesia had her baby.

I sighed. I didn't hate Lanesia, I just hated what my life had become.

I texted back questions about the baby, like the name and gender.

When my phone buzzed again, I could barely believe my eyes. Because after all the details of the delivery, I got this piece of information:

She named her boy Bentley.

Chapter Thirteen

Home Is Where the U-Haul Is

A couple of days later, I'd reached my limit. I grabbed my diaper bag, packed Tripp into his car seat, and made a decision.

Like the prodigal son, my bad choices—and sin—had caught up with me. They say the road to hell is paved with good intentions, and I know that my desire for a real dad for Tripp—a father he could count on—was a good intention. But now I was in a personal hell.

And I longed for only one thing. Home.

We all make mistakes, and while it's a cliché, it's also true that our true tests come in our responses to our mistakes. Do we make things worse? Do we stubbornly stick with bad choices even as we spiral deeper and deeper into darkness? That's what I'd done with Levi, and look where it had gotten me. This time, I was going to do the right thing, the hard thing.

And going home was the hard thing. After all, I'd seen the anger and hurt in my parents' eyes—even my grandfather was done with me. I knew that coming home was a way of admitting defeat, and I hate to admit defeat. I have a strong will, and that strong will helps me, but sometimes it hurts. That day it hurt. Part of me said, *You can do this. You and Tripp can take on the challenge together.* But the deeper part of me, the part that understands God's plan for my life, knew that sticking it out in Anchorage wasn't best for either me or Tripp.

So I got in my truck and drove back to Wasilla, the baby dozing in the backseat.

My home isn't perfect. My family isn't perfect. None are. Even on that day, I realized I'd be returning to people like Willow, who would steal my clothes, and Track, who's fiercely protective, and Dad, who calls me stubborn, and Mom, who's sometimes impatient, and Piper, who uses her own saliva as hair gel, and Trig, who is so feisty.

But they were also the people who would stick with me. Even in the midst of our problems, I knew—deep in my heart—they would answer the call. That's what we do. That's who we are.

Right?

My heart pounded in my chest and beat faster with every passing mile. The doubts began to creep in. Had they changed the key code on the gate? Would they throw me out when they saw me? Would they yell?

When I finally got there, the gate was wide open. I knew my mom's friend Juanita was there, because her vehicle was parked in the driveway under the basketball hoop where I had practiced free throws until I was perfect, or almost perfect. *That's good,* I thought. *At least I'll have a witness if Mom kills me.*

I got out of the vehicle, unfastened Tripp's car seat, and walked to the door to the shop and looked around. There were big double garage doors that allowed clearance for Dad's Piper Cub airplane. To the right was an entrance to Mom's top-floor office and to the space they'd built for my apartment.

I steeled my nerves and walked in.

Juanita and Mom were standing over near the inside steps, talking. When I opened the door, they looked up at me in surprise.

I didn't know where to start. I wanted to go back to the beginning, to explain why I'd done the things I'd done, to be understood, to be comforted. But all I said was, "Levi may have gotten Lanesia pregnant." They said nothing, but Juanita's face softened. "We're not together anymore," I continued.

Juanita immediately hugged me, patted my back, and said, "It's going to be okay."

"Why would you think you could change a guy like that?" Mom asked.

I felt smaller than a person could possibly feel as she lectured me, and I deserved every syllable. But then, after she'd gotten it all out of her system, she tilted her head and said, "Okay, we're going to get a U-Haul, and you're coming back home."

And that was that. I'm not going to say that all was immediately forgiven, and certainly all was not forgotten. This wasn't the Bible story of the prodigal son; it was a messy reunion, a tough reunion. But it *was* a reunion, and we were a family again.

Early the next morning, Dad showed up on my doorstep. Even though I'd left the comforts of their home and betrayed them, he'd driven a huge U-Haul up from Wasilla and parked right in front of my door. Behind him was every aunt, every uncle, and every cousin, no matter how distant, in my family. Grandpa and Grandma even got out of their vehicle, took one look at me, and said, "Okay, where are your dishes? We'll pack those."

Most people would've looked at what I'd done and would've changed the locks. But not this family. Not the Palins and the Heath clan. That whole day, we had a family reunion right there on my front lawn. Instead of balloons, we had packing boxes. Instead of cake, we had duct tape. Instead of remembering the past, we looked to the future.

They worked tirelessly and didn't care whether photographers were watching from the shadows. They dismantled my bed, brought down my computer desk, packed the glasses, and somehow got all of my jeans into a box. Okay, several boxes.

When the U-Haul was finally loaded, Dad jumped in it and drove straight to Wasilla.

I was headed to that imperfect house in that imperfect town. I was headed home.

Chapter Fourteen

There's Plenty of Fish in the Sea

t's hard to hide in your shame when there are cameras everywhere.

When I got home to Wasilla, all I wanted to do was hide in my Nike yoga pants with Tripp and watch television. I didn't want to *be on* television.

But I arrived right in the middle of Mom's filming of the eight-week series *Sarah Palin's Alaska*. I was a little hesitant about the whole idea. Even though they called it a "travelogue" and billed it as educational, I figured it would be "too much Palin" and "not enough Alaska." Plus, how would it be received in the chattering class? Who would even watch a show like that?

Of course, I had no idea the first episode would draw about 4.96 million viewers and be the most-watched debut in the channel's history. I just knew that I showed up in the middle of filming what seemed to be a cool series, that Mom and Dad were having amazing adventures, and that I simply wanted to tag along.

When Mom and Dad asked me to go to the Grouse Ridge shooting range—the location of Mom's baby shower when she was pregnant with Piper—I joined them. In a remark in which she spoke directly to the camera, Mom said, "The last couple of years have been hard on Bristol, because so often, what it is that she does ends up in the tabloids because of someone that she had been associated with. So Todd and I really wanted to get her away from all that and refocus on what truly matters in life. And I hope she gets that."

I'm sure that the viewing audience at home couldn't appreciate how true that statement was, or how raw all of our emotions were while filming. By the time it aired, the Levi engagement stories were old news. But when we were filming, they were current dramas. I wasn't getting any of my shots, and Mom kept giving me more unwanted advice.

"We're not gonna stop until you get one," she said.

"Mom, take your prom hair back home," I said, which cracked Dad up.

We also went south to Homer, which is the halibut capital of the world (which caused Mom to joke, "We're heading down there just for the halibut," before adding, "See, Alaskans know that joke . . . I don't know if other people will.")

Though our family commercially fishes for salmon, we'd never fished for halibut. So we jumped in the RV and headed south for a new adventure.

"I appreciate it that Bristol was game to go down to Homer with us. . . . I thought it would be good for Bristol to get away from it all, to clear her head and concentrate just on family," Mom said on the show. "All the things on the periphery that at the end of the day really don't mean anything—the things in the tabloids or the things that people make up or assume about her or our family . . . I knew it would be good to get away from all that."

On the five-hour trip, Willow and I poked fun at each other in the back of the RV. Before Mom could tell us to knock it off, the camera caught us in a typical Bristol/Willow conversation.

"We don't look anything alike," I said, speaking of my sister. "Her teeth are twice the size of mine."

"I got buck teeth?" she asked, before returning an insult. "You have no chin."

"I have no chin," I agreed, thinking it might end this line of conversation. But she piled on.

"You have big jadey eyes. I have pretty almond eyes."

"Well, you have big bushy eyebrows."

"What? I just got them waxed!"

When I saw that part of the show, which was even more hilarious before it was edited, I laughed so hard. Another part that cracked me up—one that didn't make the final edition of the show—was when Willow was looking into her phone like it was a mirror and Mom said, "Willow, quit being so vain!"

That night, we sat around the fire before an early bedtime to prepare us for a big day of fishing. But before we called it a night, Mom said, "Bristol, here's the lesson for you to think about tonight: there's plenty of fish in the sea."

That comment caused me to emit a full-throated groan, but the fact that Mom was already joking around about one of our toughest family moments was a good sign. But even though things with my family were improving, I was still haunted by Levi's ghost. We were in Homer, after all, which is where we drove together for his hockey game . . . the first time I wore his stupid camouflage coat. So, while I was staying out on the "spit," clam digging, commercial halibut fishing, and fish tendering, I was really thinking about Levi.

The TLC people billed this episode as "Todd and Sarah get a

chance to bond with Bristol," but I'm not sure anyone would ever appreciate the truth behind that bland description. While viewers may have enjoyed seeing Mom and I club that halibut, the real action was our reconnecting as a family.

Though I was initially opposed to this series, it turned out to be a great deal of fun.

And speaking of television opportunities, I said yes to the *Dancing with the Stars* offer. I printed the e-mail to show to my family, but they weren't so sure. Dad was actually opposed to my going on the show and my aunt Heather was very worried that I'd already said yes. In fact, she sent me a long text message that began:

> You've never explained why you've done all of this public stuff.

Since I love Aunt Heather so much—and always want to please her—I couldn't even finish reading the text message.

As a cast member, I'd be assigned a professional dancer with whom I'd compete against other pairs by dancing for the three judges, who each would score us on a one- to ten-point scale. Additionally, the at-home viewing audience would vote. We'd stay as long as we weren't the couple who received the lowest combined total of judges' points and audience votes.

Season eleven would air September through November in 2010. That meant I'd have to move to Los Angeles at the beginning of September and live in an apartment ABC would provide. Then I'd rehearse with a professional partner for at least twenty hours a week. The show would air live on Monday and Tuesday nights.

The only real exposure I'd had to live television consisted of

Mom's speeches and that one *Saturday Night Live* taping I'd attended. It's enough to make anyone pause, but there was a reason I said yes to the offer so quickly.

Since Tripp's birth, I was the one who had to pay for his doctors' appointments. When he grew out of his pants, I was the one who had to find the next size at Target or Old Navy . . . or better yet from hand-me-downs. And in the future when he needs braces in junior high or textbook money in college, I'll be the one he relies on. It's rare that opportunities such as *Dancing with the Stars* arise for anyone, especially a young single mom. The producers' offer was pretty generous. Rather, it was really generous. And so, I took one look at my diaper budget and excitedly said yes.

"Are you sure, Bristol?" my mom said. "This is what you want to do? Holy geez, there's going to be a lot of scrutiny."

"Mom," I said, thankful that she still wanted to protect me from the harsh light of the media, after all we'd been through, "they're going to be talking about me anyway, so I might as well."

It was enough to convince her. In fact, my dad had been discussed as a possible participant on *Dancing with the Stars* after the election, but he turned it down because he didn't have time. Not that he'd do it anyway. Having been married to my mom for so long, what kind of partner would suit him?

Mid-August, I packed my bags, loaded my truck, and prepared to drive the thirty-six hundred miles down to Los Angeles from Alaska for an adventure. (I got questioned a great deal about that, but Alaska isn't floating down there with Hawaii, contrary to how it appears on maps of the United States; you can drive from Alaska to the lower forty-eight.) I remember the day I left so vividly.

The TLC producers had gathered our whole family to be photographed in our backyard for promos. We all trudged out to our dock, warned Track not to make the funny faces he's known for in photos, and smiled.

We smiled standing together and we smiled standing apart. We smiled in the backyard, and we smiled on the dock.

While the producers and camera crew treated us like background furniture and herded us like cattle through all of the poses, I sat back and wondered what they would've thought had they known I'd be on America's top show. Maybe if I performed well enough, they would wish they'd gotten more footage of me!

My black Dodge Crew Cab was already loaded, gassed up, and ready to go. The TLC cameramen kept asking, "Where are you headed to in such a rush?"

Of course, it was a huge secret, because the show's producers reveal the participants all at once. As soon as the photo shoot was over, I kissed my parents good-bye, jumped in the truck, and headed to California.

But before I drove off with a friend for this road trip through Canada to California, my mom sang, "So she loaded up the truck and she moved to Beverly . . . Hills, that is. Swimming pools. Movie stars."

Maybe if she doesn't run for president, she could be a comedian.

What a road trip this was! We drove eight hundred miles and made it to Whitehorse, where we spent the night in a cheap hotel. After all, I didn't have that *Dancing with the Stars* money yet! The second day, we drove another eight hundred miles, where we slept in the truck! As we drove through Edmonton, I got pulled over for having tinted windows and taillights, which were illegal because they were too dark. The Canadian cop took my license and registration and looked at me squarely in the eyes. "Are you *the* Bristol Palin?"

I laughed. Since I'm definitely not a celebrity, it always surprises me to be recognized. But thankfully, this time, it had a perk—the policeman let me out of my ticket! The third night, we stayed in Calgary, Canada, and we had fun walking around in a different country, in spite of the police stopping me. The fourth night, we stayed at "a friend of a friend's" house in Bozeman, Montana. It was a total college hangout, and I felt a little awkward there. Having Tripp always made me feel so different from other people my age. Most college kids don't have a set schedule, real bills, a job, or anyone depending on them. I should've had fun there! After all, Tripp was coming down to California later, so I was free to do whatever I wanted. However, strangely it made me feel sad. I was different from everyone else, and I missed my baby. I know that Tripp is the joy of my life, and he means more to me than any rave, any superficial event, and any teenage experience.

After this stop, we headed through Idaho and then to Utah, where we stayed in a nicer hotel in a city called Cedar. When I went in to check for availability and prices, the man working the front desk asked for my information.

"Name?" he asked as he entered it into his computer. It was late and we were all tired.

"Bristol Palin," I said.

He looked up at me and tried to hide his surprise. When he asked my address, I laughed as I said, "Wasilla, Alaska." Then he finally broke into laughter, too, as he looked at my ID and back at me.

"Are you any relation to Sarah Palin?" he asked.

I nodded.

He turned out to be a really cool guy. After we went up to our room and took a shower, he called up and asked if we could pose for a photo with him. After we went out for fast food, we came back through the lobby to give him a chance at a photo, but there was also a young girl there waiting. She excitedly asked for a photo with me as well, and said, "This is the coolest thing that's ever happened to me!"

Again, it was odd and humorous to me that anyone would want a photo with the daughter of a failed vice presidential nominee. I'm not a celebrity and in Wasilla anyone can see me just by driving up to the coffee shack and ordering a latte. But in the lower forty-eight, people treated me like I was a rock star. It cracked me up, and I fell asleep that night smiling over the silliness of it all.

We finished the trip by getting up early in the morning and going through Utah, Arizona, Nevada, and then—finally—California. I was so glad to be out of that truck. And as much as I loved my friend, we were ready to get back to reality!

Chapter Fifteen

Shaking What My Momma Gave Me

My small apartment in Los Angeles was in a building where many stars live, and I geared up for a real "Hollywood" experience, though I wasn't necessarily looking forward to it. Alaskans are a very different breed of people, folks who don't care as much about appearances, clothes, and image. Plus, there aren't that many people in Alaska! Though it's by far the largest state in America by size, it has fewer than 800,000 people. That's a state that's one-fifth the size of the continental United States, though it ranks forty-seventh in population. That means that we have plenty of elbow room, and have the space to do whatever we want with a certain degree of privacy.

My L.A. apartment, however, was what real estate agents would call "cozy," but most people would call "microscopic." I'd brought a family friend to help me watch Tripp, which meant there were three of us living in that apartment. It was certainly a tight place to live!

I was paired up with a professional dancer named Mark Ballas, a ballroom dancer who'd already won *Dancing with the Stars* twice—with Olympic figure skater Kristi Yamaguchi in Season 6 and Olympic gold medal gymnast Shawn Johnson in Season 8. No pressure. When I first met Mark, I feared he'd be upset at being paired with such a newbie.

"I hope you have a lot of patience," I said, "because I don't know how to dance."

His arm was covered with Silly Bandz—little multicolored rubbery bracelets—not something you see much on the men in Alaska. Eleven-year-old girls? Yes. Men? Never. Anyway, he pulled up the Silly Bandz and showed me a tattoo on his wrist. It said, simply, "Patience," written in cool cursive.

Though I always felt like we weren't matched up well in terms of size (one of my first thoughts was "Oh great! I'm taller than he is!"), I knew the producers had matched me with the right person in terms of personality. Mark was awesome!

He wasn't frustrated because I had no experience. Instead, he just took it from the top. I honestly think he could sit there all day and explain things to people, over and over again. He didn't lose his temper, he didn't make me feel stupid, and he really made it fun. If I ever got too overwhelmed at not "getting" a move, we'd take a break.

(Wait! I can hear what you're thinking. No, we didn't date. We didn't kiss, so get that out of your mind! I love him like a brother!)

During a typical week, I'd go to the studio on Sunday for camera blocking and wardrobe fitting. Camera blocking means assigning movement of the dancers and the camera. On *Dancing with the Stars,* it meant we performed our routine onstage, and the cameramen determined where the camera should be at any particular moment in the dance. That's how the viewing audience got to see every important foot move and twirl. The cameramen know exactly what's coming, because we're putting in the hours on the weekend before the show goes live.

Believe it or not, all of the wardrobe is made specially for each contestant. Everyone was wondering how I'd be able to pull off the costumes, some of which barely cover your . . . assets. I've never been the type of girl who shows cleavage or stomach, so I told them from the get-go that I wasn't going to be wearing anything sleazy.

"I'm not going to show my stomach because I have a son at home," I told them. "And I'm not going to show cleavage because that looks just like a butt crack."

Some of the stylists got frustrated with my restrictions, but they ended up making me some very pretty costumes, all of which were high necked and as modest as you can get on a dance show.

After weeks of practice, the first night rapidly approached. You're only allowed so many tickets for visitors each show. On the first night, all the contestants negotiate to get as many people in their families as possible in the audience there to support them. I worried about getting enough tickets for all the family who'd want to come. After all, through the ups and downs of the weird spotlights that we had found ourselves in—from middle school football to the GOP convention to numerous Iron Dog competitions to this—we were always there for one another. As we talked about how they were coming down, I begged my family not to drive our motor home down here. (Can you say Wasillibillies?) The producer kindly granted me tickets for all of my family except Track, who handles public appearances like Superman handles Kryptonite.

To be honest, I was not looking forward to going out on that stage. My mom wasn't too popular in Los Angeles, and I was worried the crowd wouldn't necessarily be in my corner. After being away from Alaska for what seemed like so long, I couldn't wait to look out in the audience and see the supporting faces of my family.

On the Sunday after we did our blocking, however, Mom and I had a talk about them coming to the first night of performance.

"Bristol, I thought you knew," she said. "Remember, I had that previous commitment?"

I was devastated. Though I understood how busy she was, she's still Mom. And sometimes a girl just needs her momma, no matter how old you get. On Monday morning, I went into the studio for yet another camera blocking, dress rehearsal, and wardrobe fitting. And there, in the seats of the empty studio, I saw my family's names attached to the chairs still reserved for them.

Todd Palin, Sarah Palin, Willow Palin, and Piper Palin.

I knew having my mom there would be good for the *Dancing with the Stars'* ratings, because the press had already been buzzing about whether she'd show up for what some critics called a frivolous show.

I had to break the news to my boss Deena that my parents weren't coming and were instead having a *Dancing with the Stars* party in our Wasilla living room. However, she didn't act disappointed at all. In fact, she comforted me. I think she could see my disappointment.

"It's okay," she said, putting her arm around me. "I bet they'll come soon."

Though her kindness showed how much she'd taken me under her wing, her words missed the mark. After all, I had no idea how long I'd last on that show. My main goal was to survive the first week. This was my moment—my big, nerve-racking moment—and I wasn't sure how many more of these "moments" my two left feet would provide. Thankfully, however, I had Mark. He treated me like a princess and was an amazing instructor. Without Mark (and his mom and dad), I wouldn't have had such an enjoyable experience and I'll always be grateful for his friendship.

On our first night of competition, I took a deep breath, steeled my nerves, and walked out on the stage with Mark. When the cameras went through the audience, the only people sitting in the seats formerly labeled "Palin" were Dr. Cusack's kids, who came to fill the seats as a favor. (I was so thankful that my boss's kids jumped on a plane Monday morning to be there with me.) I had to take this journey—at least the beginning dance steps of it—alone.

Actually, I did have help. Back in Anchorage, when I got a full-time job, I hired a person who lives in Anchorage to help me take care of Tripp. This was the first time that I had help with him, and it was hard for me to leave him in the care of someone else. However, I found the most perfect caretaker who is so close to him that she now even considers Tripp her own grandson. I couldn't have worked at Dr. Cusack's without her loving care of my son. And, especially, I couldn't have done *Dancing with the Stars* without her.

We were assigned "Momma Told Me (Not to Come)" as our song because it sadly summed up some of my rather embarrassing personal decisions. That night, I knew it was important to make a big splash. And so I wore a very conservative gray suit jacket and skirt. Oh, and, of course, I wore an American flag pin on my lapel. But just a few seconds into the dance, I ripped off the conservative costume to reveal a red-fringed minidress underneath. The judges gave me a combined score of 18, which wasn't too bad for a girl from Wasilla who'd never danced in her life.

Who needs the prom?

Would it be enough for me to make it through the first episode? During the shows, the judges' scores didn't determine our fate alone. People watching at home could call, text, and vote in online polls to help get their favorite to the next round. The couple with the lowest combined score of all those factors gets voted off the island. Oops, that's the wrong reality show. The one that gets the fewest votes gets stuck with a $90,000 bill for vice presidential "vetting." Oops, that's the wrong thing, too. The one with the lowest score is off the show, but the producers said we could stick around until the last episode, when all the contestants come back for the grand finale.

I actually wanted to make it to the last episode, though it sounded overly ambitious to say so aloud. The show pays contestants increasingly more money each week, and I wanted to dance my way into getting enough money for real estate investments so I could provide a great college fund for Tripp.

Amazingly, I made it through the first week! In a rather shocking twist, a nice guy named David Hasselhoff (whom I'd never heard of before this show but apparently drove a talking car in the 1980s and is still a hit in Europe) was sent packing. Though everyone was shocked at his departure, I lived to see another week and accomplished my meager goal.

However, there were things in Alaska that needed tending to—bills, a dentist appointment, and other loose ends. I told Deena that I needed to make a quick trip to Alaska during the week, and she decided to send Mark with me! He'd never been to my wonderful state, so he was pumped as we headed to the airport at 6:00 A.M. Wednesday morning for a flight. (It was a faster trip than my five-day truck trip.) Then we drove to Sonja's Dance Studio (Wasilla's only one!), where Piper and my cousins learn hip-hop, jazz, ballet, and tap dance.

When we arrived, hundreds of little girls were all lined up to welcome us. They had signs that read Shake What Your Momma Gave Ya; Dance, Bristol, Dance!; and Team Palin. Also, after the difficult presidential election and the tabloid Levi drama, it just felt wonderful to be doing something that had nothing to do with politics or scandal. The entire town seemed to be behind us, and it was wonderful to show Mark all of that small-town support. We rehearsed a few nights right there in Wasilla.

Mark said he just wanted to see a moose, so he went to a wilderness park (where he got within five feet of a grizzly bear) and got to see porcupine, caribou, eagles, and bison in their native habitat. Yes, and moose. He also got to fly in a seaplane and see glaciers from above!

I also got to show him my home, where Mark got to meet my mom, my dad, and Track. Mom was so happy to meet him because she was thankful I'd been paired up with such a patient gentleman. She also, in one mortifying moment while cameras rolled, imitated my "shimmying" move and asked, "How did you teach her to do that?"

I thought I was going to die! It just goes to show that all teenagers are sometimes embarrassed by their moms, no matter how famous their moms are.

It was a whirlwind trip. Mark and I hopped on the redeye back to Los Angeles so we could rehearse on Friday and over the weekend. On Monday we were to perform the quickstep, and

I was a little nervous. There was a part of the routine that I just couldn't get right no matter how hard I tried! The trip—though fun—had taken up valuable rehearsal time.

When it came time to perform, everything began well. The team of Rick Fox and Cheryl Burke did the jive, then Florence Henderson gracefully danced the quickstep with her partner, Corky Ballas. Notice the last name? Florence's partner was actually my partner Mark's dad! (It was fun to be with Mark and his dad on the same stage. They got along so well, and Corky kept my mom up to speed on *Dancing with the Stars* progress via e-mails.) When judge Bruno Tonioli said her performance reminded him of *Driving Miss Daisy,* everyone booed. Don't mess with Carol Brady, Bruno!

But there was more booing to be done that night. A few contestants later, Jennifer Grey and her partner, Derek Hough, performed the jive—and the audience ate it up. At the end of their very good dance, they both collapsed on the floor to show they'd given it their all. The audience rose to its feet and applauded. Like always, they made their way from the judges to cohost Brooke Burke, who was waiting for their postdance interview. But when their scores were announced—solid 8s!—a loud boo came from the audience.

It was so unexpected, it was impossible to ignore. So, on live television, Jennifer awkwardly turned to Brooke and asked, "Why are they booing? What's the booing?"

Brooke obviously had no idea. The cameras quickly shot back to cohost Tom Bergeron, who by this time had made his way to where my mom was sitting in the front row. This made it look like the boos were somehow directed at Mom.

When she was first on camera, Mom was shaking her head—agreeing with the audience's reaction to Jennifer's low scores. Tom went on as planned, asking Mom some softball questions before asking her what she thought of the judges. This is where all of Mom's skills honed on the campaign trail came into play.

"It's like before a hockey game, you're not going to chew out the refs before your team is up! So you're great," she said, turning to the judges. "You guys are doing great! Bristol the Pistol, that's who we're rooting for!"

Well played, Mom.

As soon as my mom's interview aired, blogs lit up all over the Internet.

"Was Sarah Palin just booed on *DWTS*?"

Even *Time* magazine—the hard-hitting news mag that it is— weighed in on the subject, claiming that the Los Angeles audience wasn't too fond of "Momma Grizzly." Typical mainstream media . . . always thinking the worst!

Tom Bergeron later came to our defense and told the public what everyone in the studio audience already knew.

He explained, "Last night, the booing heard was a direct response from the audience to Jennifer and Derek's scores, which were perceived as relatively low." He also explained it to the media later to straighten out the issue. He even called it "boo-gate."

It's funny how nothing my mom does can escape criticism!

We were the last to perform, topping off a night that contained the infamous Michael-Bolton-in-a-dog-collar performance. I was nervous as we waited our turn; I just never could get part of the steps down. My nerves were calmed just knowing that Mom, Willow, and Piper were out in the audience grinning from ear to ear! When I went out there, I had no fear. And guess what? I nailed it!

Though I still didn't love dancing, I felt like I was getting the hang of it. The next week, Mark said that I was starting to think like a dancer, instead of like someone learning dance moves. I thought that was a nice compliment . . . which I needed to hear to get me through all of the work!

Every Friday through Monday, we worked at least fourteen-hour days. There was always something else we needed to practice, a costume that needed hemming, a step that needed to be mastered, a photo that needed to be shot. We even filmed our dances so I could watch the moves on my iPhone when I wasn't in the studio! It was almost as hard as being out there fishing with Dad in Dillingham. (No, actually, it wasn't even close!) Mark and I began working on my facial expressions because judge Bruno had said I wasn't dancing with my whole body. So I practiced smiling and "emoting" along with my dance that week.

Talk about getting in character! Mark and I decided to wear gorilla suits for our performance! Everyone was laughing at us, because we were having a great time when they did our "backstage shots"—jumping on couches, beating our chests, and generally "monkeying around." I think the other dancers thought we were ridiculous, but at least we wouldn't be forgotten.

Lauden, Willow, Heather, and Juanita's daughter Jenna all came down for this performance because it was my twentieth birthday. When we did get the lowest scores of the night—just 18 points!—I didn't care at all. I'd butchered the steps, but so what? I was surrounded by so many good people that the competition didn't matter. (It was similar to the warmth I felt at the GOP convention, when I was able to withstand the chaos because of my family's presence.)

A few days prior to that night, Mark had surprised me for a little pre-birthday fun. He knew my family loved hockey, so he secretly decided to try to get me tickets to an L.A. Kings hockey game. After calling around, he scored front-row tickets! I was so surprised. He didn't tell me where we were going, but we drove into the parking lot of the Staples Center. I couldn't stop smiling! Both Mark and I were given custom jerseys, with our last names on the back, and they put a birthday greeting on the big screen for me! The enormous players skated so well on the ice. Since we don't have a professional hockey team in Alaska, it was a treat for me to go to an NHL game. Plus, we met Vince Vaughn, who sat near us, and a couple of the players after the game. It was so cool to meet him!

Also, someone else gave me tickets to an awesome Blake Shelton concert at Club Nokia that was part of his All About To-night Tour. The crowd sang along to his songs and cheered when he said, "You don't know how nervous I've been about performing here in L.A. I'm just glad that there's some hillbillies like me out here!"

I loved it! I'd been in Los Angeles for so long that Blake Shelton was a breath of fresh air. I always listen to country music, and I felt that he was totally normal in a city of people made of plastic. The people I met in California were so obsessed with their bodies, their clothes, and their cars. They'd talk about getting silicone injections to make their butts look bigger or liposuction to make them smaller. They had philosophies about how to have the best nails or the poutiest lips. They'd talk about the latest fashions and compare the newest Mercedes and BMW styles with much fervor and disagreement. I could never really weigh in on the controversy, though. I kept my opinions on the Dodge versus Chevy truck debate to myself. There's nothing necessarily wrong about the way they lived, but being consumed with image is just not something I'm used to. I always related more easily to the receptionists than to the pro athletes and Hollywood starlets.

So the country music concert was so much fun for me. I was in the front row, and Blake Shelton winked at me when he sang. (I laughed when the two guys beside me were thrilled because they thought he'd winked at them!) I also got to go backstage and meet him and was so touched when he called me "America's sweetheart."

"Well, your album was the first one I bought on iTunes," I told him.

Mark had been so nice to me that I didn't expect even more on my actual birthday! However, after our low-scored performance, Mark, some friends, and I went to a cool place called the Mint. Mark had a cupcake delivered to me, with a candle and flowers, and he sang "Happy Birthday" to me from the stage. It was a sweet moment, a nice ending to a difficult week performance-wise. Plus, it felt kind of good to say good-bye to my teen years! (I was officially no longer a "teen mom"—something to celebrate.) Erika from Juneau came to that show and it was so nice to see her. In fact, the whole experience was a wonderful reunion, something my friends and family were able to gather around, which was pure fun!

Sadly, that week Florence Henderson and Mark's dad got sent home. Florence was such a wonderful "mom" figure for everyone. She remembered everyone's name, including production assistants, producers, and cameramen. She was so kind with everyone, and it was no surprise that she delivered a very gracious send-off speech when she was voted off. Mark was disappointed to see his dad go, because they so enjoyed hanging out backstage and going through rehearsals together. However, her departure was less terrible because she was always in the audience afterward, cheering us on.

Week after week, Mark and I made it through, and we lost

friends who'd been voted off. Plus, the schedule got even more intense. The longer you last in the competition, the harder you have to work because you have to learn more dances. Since there are fewer dance teams left, we had to start doing even more media interviews. One particular day, I had rehearsal from nine in the morning until five in the evening. Then I shot a public service announcement from six until ten at night. From ten until midnight I did a photo shoot with *In Touch*.

Oh, and speaking of the public service announcement, I filmed that with Michael Sorrentino, publicly known as "The Situation," from his hit MTV show *Jersey Shore*.

Because of his filming schedule, he arrived a couple of weeks later than everyone else. Honestly, I was worried. Would his personality throw everyone's balance off? Would he be abrasive and rude? Would he constantly be referring to himself in the third person?

However, nothing could've been further from the truth. We got along so well, in spite of our different backgrounds. I also found it interesting that he was so business-minded. While it might not shine through on *Jersey Shore*, he is always thinking about and strategizing about business ideas.

When the Candie's Foundation called and suggested Mike and I do a commercial together, everyone thought it was an odd pairing. I told my mom I was going to do it, and she thought it was hilarious. (I thought it was hilarious that Mom knew who Sitch was!)

You may have seen it on television or online. But it starts with him checking me out before he realizes it's me.

"Excuse me, miss," he says. "Have you ever had a situation with the official situation?"

I think the word *situation* got used so much in this thirty-second PSA that it should've probably been retired afterward.

"I hope you're as committed to safe sex as you are to those abs," I said to him.

"Just in case you do get into a situation, I want to make sure you are situated. Because if you do get into a situation, with your situation, you may end up with a situation."

"Trust me, I'm not getting myself into another situation," I said. This commercial had only a skeleton of a script. The rest we ad-libbed, which is why at one point, we exchanged a few too many "for reals."

For real?

For real, for real.

Anyway, Sitch got some silly lines in. For example, he said, "I totally respect abstinence. I mean, it actually has the word *abs* in it!" When I asked if he practices safe sex, he said, "I practice all the time!"

Many people criticized me for being in a commercial with this guy, but I thought it was exactly the right way to reach an audience not used to hearing about the option of waiting until marriage for sexual activity, and how difficult it is to be a teen mom. After all, you don't want all of this conversation to be aimed at church youth groups!

It was actually one of the most fun times I had while being a Candie's Foundation spokesperson.

So, on top of all the other *DWTS* work, this PSA was actually shot during our season. Again, there were lots of publicity shots, camera blocking, and wardrobe fittings, but not too many really intense workout rehearsals. Oh, and we had lots and lots of hair, makeup, and spray tanning—we even had "body makeup"! I'd have to sit in that chair for hours upon hours, which gave me a lot of time to satisfy my Craigslist addiction for real estate deals and trucks. If you are one of the few people who haven't discovered its joys, it's an online network of free ads. If you need a job, a car, or someone to teach you Spanish, it's a good place to start. Since the hair and makeup sessions were so frequent—and long—I'd frequently pull up Craigslist on my iPhone and shop around. I could tell you the price of any used vehicle in Alaska by the time *Dancing with the Stars* had finished wrapping. It was then that I saw a nice house for sale in Maricopa, Arizona. I didn't have any real connection to that state; I just knew its housing market had tanked, and my desire to find a good deal kept me prowling around the real estate ads. I bookmarked the house and hoped I'd be able to stay on the show long enough to afford it. Surfing Craigslist—though potentially expensive for someone like me—allowed me to escape from watching negative coverage of me on the show. Because I didn't read what people were saying about me, I was surprised to find out that my continued success in the show was causing a bit of a stir.

First of all, people complained that my mother's "Tea Party supporters" had organized some sort of conspiracy to help keep me in the competition. The speculation had gotten so out of hand that legitimate news organizations (and I use the word *legitimate* loosely) began doing reports about the mysterious phenomenon that was keeping me on the show week after week, in spite of (ahem!) comparatively low ratings from the judges.

In fact, executive producer Conrad Green had to go on record defending the show's voting process and said he was mystified that people were complaining so loudly about my continued participation. I tried not to take it personally. I'd really improved over the course of the show, and no one was giving me credit for it! Well, no one except the people who counted . . . voters who kept calling in their support. (Thanks, by the way, to all who called!)

But the controversy didn't die down. One man in Wisconsin was so angry I'd advanced to the next level that he actually screamed "that's f—ing politics!" before shooting his television with his shotgun. Then, he turned the gun on his wife, sparking a fifteen-hour standoff with police. I read about that and told my mom I'd buy his wife a new television set.

Then I heard from other "fans." A letter was sent to the production offices addressed to me, but instead of a note of adoration, inside there was white powder. Police, firefighters, and the FBI showed up at the studios to investigate the "death threat" against me, but determined the substance was merely talcum powder.

Even though I never was rated too highly, I didn't take the criticism from the judges personally. I figured there was so much bad stuff on the Internet about me, what could they say that was worse? While some contestants talked back to the judges and were devastated at the slightest critiques, I just took it with a grain of salt. Anyone who's rocked a cranky toddler to bed at three o'clock in the morning or wondered how they'd pay a bill knew this competition wasn't real life.

However, it was hard not to take the tension among the contestants personally. At first, no one looked at me as a threat, and everyone was really kind. However, as each week passed, people got a little colder. I noticed some of the contestants rolled their eyes when they realized we'd survived to dance another day. A lady in wardrobe got me the wrong size Spanx (an undergarment that helps you look your best under costumes that leave little to the imagination) and went to get the right size. While I stood in the makeshift "changing room," a temporary structure that allowed us to change behind paper-thin walls, I heard her say, "Bristol's such a b—ch."

As luck would have it, Mark came in at the exact time and heard her comment. The ever-protective older brother, he confronted her. "What did you just say about my partner?"

Even though Mark was friends with everybody on the competition, relationships began to strain when they were voted off before we were. They literally would not speak to us—or even acknowledge our presence!

All of this over a dance competition and a pretend disco ball trophy?

At the same time that a poor television in Wisconsin met its demise and the white powder insinuated mine, there was even more controversy.

Apparently, a comedian with very little material decided to get media attention by using me as a punch line in her stand-up routine. She said that I was the only person in the show's history who actually gained weight over its course. She even called me the "white Precious," a reference to a movie about an obese African American girl impregnated by her father.

The ironic thing about this whole thing is that my critic was a self-described public activist against bullying . . . as long as she agrees with the victims' politics, I guess. After a prominent gay suicide, she went to the airwaves and criticized bullies . . . while at the same time publicly stating that a 135-pound young girl like me was obese. But because I was already active and fit, the show's schedule wasn't new to me. I did gain five pounds over the course of the show, but I was still thinner than a lot of girls my age. I'm not saying this because I'm somehow proud of how I look. Rather, I'm trying to give perspective on this fifty-year-old woman's criticism of me. When the public didn't think her gibes were funny, she defended herself by pointing out that there was a big difference between being a bully and a comedienne. Apparently, it's okay to make fun of impressionable teenagers—and to spread the lie that a healthy weight should be mocked—as long as it's profitable. (I was in good company, however. She also called another prominent Republican's daughters "prostitutes." Why aren't you laughing? Don't you get the joke? They're *Republicans* and they're *young girls.* Isn't that funny? Apparently, you don't have a good sense of humor.)

So let me address this issue head-on. I'm not skinny and I'm not fat. I'm a girl like everyone else who lives in this era of airbrushed photos who's trying to maintain a healthy lifestyle. And do you know what? I'm succeeding. Everyone talks about how strenuous the *Dancing with the Stars* workouts are, but they're nothing compared to any high school basketball practice. Though one contestant famously lost forty-one pounds, I have to assume it was because she was not too active beforehand. Because I was an active kid before I started, this show actually took my activity level down a notch. That meant I was not going to be seeing rapid weight loss. Which was fine. After all, *The Biggest Loser* is on a different network altogether. We were doing a dance competition, which meant tights and stretchy pants every day. Like most women, I can tell how much my weight has fluctuated by which jeans are too snug in the waist. Since I wasn't wearing many jeans, I didn't realize I'd gained five pounds. But the rest of America sure did. When I realized people were criticizing my weight, I made a vow. I decided I was going to go to the end in *Dancing with the Stars,* even if I wasn't the skinniest or the hottest girl on the stage. Also, I threw down a little trash talk to make it interesting. "Going out there and winning this would mean a lot," I said in one preshow interview. "It would be like a big middle finger to all the people out there who hate my mom and hate me."

And speaking of my mom, she ended up coming to several shows. In one memorable moment, I told her I had to dance the *paso doble,* and I could tell she was impressed at how easily the words rolled off my lips.

"How on earth can you dance the *paso doble* when it's going to be hard to even pronounce it?"

That's when Piper said, without missing a beat, "I know how she can learn the steps, Mom." Then she turned to me. "Bristol, just write it on your hand."

Believe it or not, Mark and I got a standing ovation! My mom, my dad, and Willow were in the audience, and they clapped excitedly as I got my highest scores of the season. The judges were very kind to me.

Judge Carrie Ann gushed, "This is what we've been asking for all season long, for you to come out and nail it!" (Even before I went out onstage, Maks came up to me and whispered quietly, "I know you don't like me, but good luck.")

For our second dance, Mark and I discussed what song would really blow them away. I wanted to dance to Gretchen Wilson's "Redneck Woman." It would've been hilarious, and the crowd would've eaten it up.

Mark, wisely, wouldn't let me do it. "We aren't doing any of that awful country two-step stuff." (And think about Mom's reaction? It almost would've been worth doing just to see her melt into her seat!)

So we discussed it and tried to find something that really— that actually!—captured my essence instead of the stereotype people *think* I am. That's when we decided to do a waltz to a classical piece from Mel Gibson's film *The Passion of the Christ*. I wore a black dress and came out onto the dance floor in a hood. The music was dramatic, and the performance felt very emotional to me. Mark (who's Catholic) and I thought it was kind of fun to sneak in that song since the show usually showcases songs that are . . . well, a little less inspired.

In the trailer on Mark's iPhone before we went out, we watched the scene from the movie where Jesus is carrying the cross and Mary is watching him do it, and she has flashbacks about him as a kid. The dance came together so easily. The slow waltz . . . it should've been the hardest one. . . .

Making it to the end of the show was implausible, of course, since I didn't know the cha-cha-cha from the rumba. Yet, on November 16, I was backstage, awaiting my fate. Would we get voted off, propelling pop sensation Brandy and her partner, Maksim, into the final week? Or would we make it to the finals?

Right before we went out onstage, Brandy grabbed my hand and said, "You know, this has been so fun with you." I could tell she (along with everyone else) believed I was going home and was trying to make me feel better about the end of my journey. However, for some reason, I believed in my gut we were safe.

"You know what," I said. "This is in God's hands. Whatever happens, happens."

"Yes," she agreed. "It's in God's hands. We better stay in touch, BP."

When we were finally standing onstage, the lights were dimmed in the studio as we stood there awaiting our fate. The anxiety-producing music—which sounds like a beating heart—began playing.

Host Tom Bergeron drew out the announcement as long as possible.

"Brandy and Maks," he began. "Bristol and Mark. On this ninth week of competition, I can now reveal that the couple who received the lowest number of votes was . . ."

At this point, the host apparently decided to give his voice a rest. He may have jumped backstage to read *War and Peace*. Or maybe he decided to take a nap. But when he finally got around to announcing who won, all chaos broke loose.

"Brandy and Maks," he finally said.

Immediately, the cameraman took shots of obviously shocked contestants who assumed we'd be the ones going home. At home you may have seen Brandy's shocked expression and heard the audience gasp in horror. However, you didn't have the privilege of hearing Brandy's mom jump up and start yelling at the perceived injustice, right in the same area of the crowd as my mother. "This is rigged!" she yelled. "This is rigged!" Several of her friends and family stood up and started yelling, causing a miniuproar in the crowd. Immediately, when I saw the obvious consternation of the people in the room, I started crying. Mark attempted to comfort me, but it was hard to shake the feeling that this competition was honestly not worth it and that people had their priorities so screwed up.

Right then, I was ready to go back to Alaska, where the weather's cold, but most people aren't. However, after calming the audience down, the show's casting director, Deena; Mark's mom; and my mom pulled Mark and me aside and gave us a pep talk.

"Don't let this ruin your moment," Mark's mom said.

Jennifer Grey and Lacey Schwimmer calmed me down enough so that I actually wanted to go to the finals.

My mom chimed in, "You guys just made it to the finals, when no one said you could!"

It took us a while to shake it off. The last week of re-hearsals—which should've been a great deal of fun—was tainted by all of the controversies of the season. It was also weird to do camera blockings, rehearsals, and wardrobe fittings with only three couples left.

Before the show, Mark's managers, my parents, my cousins Brandy and Greg, and I went to Mark's trailer to chat and try to calm our nerves. There was something about being there just re-laxing and laughing about the highs and lows of the season that felt rejuvenating. When it was time to take the stage, however, the pressure returned with a vengeance. For the finale, we had—count 'em—*four* dances! That meant we had to choreograph and rehearse four times as much as we did in the beginning . . . and in only five days!

"Whatever the outcome is," I told Mark, "I'm just excited we got this far."

Our first challenge was to re-dance our most favorite ballroom dance of the season, and we chose the tango. I realized how far I'd come as I felt so much more confident this time than when I'd performed it last time. However, when the judges revealed their scores, we were still on the bottom.

Kyle Massey and Lacey Schwimmer got 26, Jennifer Grey and Derek Hough got 30, and Mark and I got 25. No matter. I felt the score was expected and was happy at my improvement.

The last dance of the season was really challenging. The judges wanted us to dance the cha-cha. I was so glad this journey was about to come to an end, and I decided to have as much fun as possible out there onstage.

Right before we went out onstage, Tony Dovolani, a professional dancer, helped me a lot and was very encouraging. At the end of my dance, I looked at him in the audience and stuck my tongue out at him. He looked back at me and said, "That was awesome!" The judges even said it was my best performance of the season! Kyle and I laughed that we could easily predict the order of placement before we even performed. And when the scores were read, we were—say it with me—at the bottom. Kyle and Lacy got 28, Jennifer and Derek got 28, and we got 27.

"We'll now reveal that the couple in third place is . . ." When it was time for America to find out the finalists, I was completely relaxed. And even when he announced our names, I was fine with the outcome.

It had been a difficult three months, but we ended it with such fun and joy!

Mark immediately hugged me, and the audience rose to its feet as we went down to talk to the judges. The host, Tom, came up to us, and I realized then just how much I appreciated him. He'd told me previously that he was the most liberal of Democrats, but that he could tell I was putting my all into this competition. Right after we received the news, he summed up how I felt pretty well, when he kindly said, "All the other nonsense aside, this is a girl who got in her truck, drove five days, and made it until the last night of *Dancing with the Stars.*"

That's when the producers did a montage of my experience on the show. They took us all the way to the beginning—when I'd just met Mark—and then showed all the highlights of the season—

including my rather embarrassing first attempt at "shimmying." (At least they didn't show my mom's attempt!) When I looked over and saw Mark cry, I teared up, too!

Even though we came in third, I'd managed to accomplish both my spoken—and unspoken—goals. I lasted through the first week, and I lasted as long as anyone else. That meant, I collected as much money as the winners (though Jennifer did get the mirrored ball), and I was able to buy that investment home in Arizona!

But do you know who else won? The controversy surrounding our survival brought more than twenty-four million viewers to the results show. It was *DWTS*'s biggest audience in six seasons.

"Wow!" my mom said. "Keith Olbermann had 200,000 viewers and you had 24 million. Not bad!"

After my *Dancing with the Stars* appearance, I got a taste of the leftist media's treatment of people they are intolerant of when MSNBC named me one of the "Worst Persons in the World" on Olbermann's show. Since then he's been axed, has seemed to disappear, and is more irrelevant than ever.

But it wasn't over quite yet. The show ended on Tuesday, so we flew through the night on a private jet. On Wednesday morning, the final three couples made our final appearances on *Good Morning America*. On the show, I danced one more time with Mark, and then the producers surprised us with personal video messages from our families.

"You overcame a whole lot of challenges starting from ground zero to come so far," my mom said on my video. "All of Alaska, we're proud of you. Way to go, Bristol the Pistol! We're proud of you."

I teared up. There was something about seeing Mom's face

that made me feel unusually emotional. We'd been through a lot over the past few years, and we'd managed to end up stronger than when we started—through repentance, forgiveness, and the unconditional love of family.

The privilege of seeing this play out over the course of twelve weeks was what I really won on *Dancing with the Stars*.

Chapter Sixteen

Seeing Things Clearly

After *Dancing with the Stars,* I knew more about "body makeup," *paso doble,* and David Hasselhoff than I ever wanted to know. I left with great friends, newfound confidence, and a hefty check that I used to invest in real estate to help secure my future with Tripp.

But none of it was "real." Though emotions sometimes got carried away, they were the kind of emotions that spring from unusually cool circumstances, glimmery disco balls, and lots of television cameras. The producers made some pretty poignant "backstory" segments to add emotional interest while the contestants rehearsed. That's where viewers learned that Kyle Massey's family was just about to move back to Georgia when he finally got work on *That's So Raven,* about Jennifer Grey's automobile accident, and about Florence Henderson's grief over her husband's death. Though the segments were sleekly produced tearjerkers, they at least hinted at the "reality" that this "reality television show" had taken us away from. The backstories, were, of course, the real stories.

When the show was over, we all faded back into life and *Dancing with the Stars* became just a chapter in our lives. (In my case, quite literally!) I was happy to return to my normal life, to help forge a new "backstory" for myself into something better than being a "teen mom" or "teen activist." After all, I'm twenty years old now!

I relaxed in Arizona, where I had just purchased my new home with my earnings from *Dancing with the Stars*, after the show before I returned to Alaska for Christmas. I was thankful to get back to a simpler life—the kind that consisted of just Tripp and me and snow-machines and absolutely no spinning around in heels and sequins.

One of the cool things about my "backstory life" is that sometimes I get wonderful opportunities to have eye-opening adventures. *Dancing with the Stars* was one of those experiences, but I had a chance at another when Mom invited me to tag along with her, Dad, Greta van Susteren and her husband, John, and the Reverend Franklin Graham on a humanitarian trip to Haiti. Reverend Graham's organization is called Samaritan's Purse, and they were going to help provide relief for the country that had been ravaged by a terrible earthquake at the beginning of the previous year and ravaged by a cholera epidemic at the end of it. When Mom, Dad, and I had the opportunity to see this horrible epidemic up close, I wasn't sure how I'd handle it.

However, I packed my bag and got on a plane. I definitely couldn't pack my truck and drive there!

By the time we arrived, thousands of people had already died. Millions of Haitians were living in tents and huts along the streets and in the rubble in unbelievable conditions. At the cholera treatment center, I saw an eight-month-old boy whose mother had just passed away. The baby's father was caring for him and six other children. He'd walked nine hours through the night from his hut up in the mountains to get the little baby boy to the treatment center. He had nowhere else to go, no other hope, and he barely made it. The baby had so many IVs in his tiny hands his whole body was beginning to swell, and his baby feet—so full of the IV solution—were rock solid. The father was talking to the translator. Though I couldn't understand what he was saying, I could tell by his emotion and urgency that he was trying to give the baby up for a chance at life.

Though they didn't think he'd live through the night, God gave him life.

I actually offered to take the baby home, but you obviously can't just pick up a child and take him out of the country! (It felt wrong to leave him there, though, whatever "the system.") I don't know what will happen to that innocent baby boy, but I will never forget him—or the look of desperation on his father's face.

On our second day of the trip we went to another treatment center filled with cots. Staff had cut holes in the middle of the cots to use as toilets because patients couldn't control their bowel movements. Also, they continuously vomited so that everything in their system was depleted. That's what cholera does; it just dehydrates you to the point of death. There was no dignity to be found anywhere in the tent and very little hope. We couldn't do anything at all but pray.

After we walked back to the children's ward, I met identical twin boys who weighed about five pounds each, were dressed in girls' dresses because they had nothing else, and looked to be only a week or two old. I was astonished to find out they were six months old. Their mother died right after their birth, so their grandmother had walked with them to the treatment center, but she could no longer care for them. She walked around and asked everyone if they could take them. I would've taken them. I should've.

Next we got to pass out the Christmas shoeboxes that Samaritan's Purse collects from people all across America. If you're one of those angels who pack a shoebox for that organization, please know that the boxes are appreciated. When I handed the inexpensive boxes to the children, their eyes lit up like an Alaskan kid's eyes might if he'd received a new snowmachine.

It really gave me perspective.

All of the frustration I felt over the rude speculation about my *Dancing with the Stars* weight disappeared as I watched people without food. "Body image" problems only exist because of our country's wealth, our prosperity, our laptops connecting us with blog accounts, those pesky cameras that add fifteen pounds, and those airbrushed magazines that take off thirty.

After I returned to Alaska from our trip to Haiti, I spent a lot of time reflecting on that experience, thinking about its implications for my life.

Some people may question why I'd be willing to be so honest and candid in this book about my teenage mistakes and problems. (Including my parents, who'd probably rather know fewer of these details!) In that stupid *Us Weekly* article about our engagement, part of my public defense of Levi to my mom and dad included this sentence:

> If a mistake is made the honorable thing to do is to own
> up to it.

Of course, I was talking about Levi owning up to *his* mistakes. But as I finish this book, I realize it applies equally to myself, and to all of us as we try but fail to live up to standards we know are right. Our inability and unwillingness to keep the standards don't make the standards any less valuable and good.

Throughout my life, I've learned that sexual standards are vitally important for a person's sense of well-being. The connection between sexual restraint and emotional stability for girls and women is especially important—I noticed that most acutely when I was at home with the baby while Levi was living his normal life uninterrupted. That's why it's important to wait until marriage to have sex, which guarantees that babies will be born into actual families instead of the patched-up kinds that we try to make work between custody and shared holidays. (Or, in my case, all alone.) The happiest people are those who live as closely as possible to the biblical standards God laid out for us . . . even if you've already violated your own sexual principles.

One of the reasons I shared this story is to convey a simple truth to the teens who are out there reading this book. (Or maybe even some adults!) If you have made a sexual mistake, you don't have to fully give in to that sin. You can always choose to live by biblical standards, which means—among other things—not having sex if you're not married and not having affairs if you are.

I've not always "walked the walk" when it comes to standards. That's the thing that struck me in Haiti. The amazing volunteers we met there definitely talked the talk *and* walked the walk. They worked tirelessly for hours, sacrificing their own health to save those who were suffering. Seeing these dedicated volunteers made me recommit to living for God and serving others. I also found myself wishing, like so many people, that I had even more of a clarified true calling . . . something I was so passionate about I'd sacrifice everything. Maybe that will come soon. Maybe it will be somehow related to poverty in other countries, or to orphans who need homes. Maybe I'll go into politics, maybe I'll write children's books, maybe work at Dr. Cusack's office, or maybe I'll help my mom become the first woman president!

While I don't know what the future holds, I do know I've made a decision not to have sex again until I'm married. (I've had one partner in my entire life, and that's one partner too many for an unmarried twenty-year-old.) That decision will help me try to live out my dreams and find my way in this world without being burdened by the bad decisions that previously haunted me.

Let's face it. Making mistakes and dealing with them, suffering pain and longing for a better day . . . that's just all a part of life. While I was finishing up this book, I got some terrible news about Hunter Wolfe—the first boy to send me flowers, the guy who'd leave me notes on my windshield during basketball practice, and who enjoyed gourmet grilled cheese sandwiches at the Governor's Mansion.

He had committed suicide.

Mom told me the first time someone close to you dies is the hardest, that, unfortunately, in an odd way, I'd stop reeling from shock as I age and get more used to the tragedies of life. While I know that's true, I just felt so devastated that a kid with such kindness and potential would end his life in such a hopeless condition.

That's one of the reasons I decided to be honest and candid in this book. Everyone wrestles with the indignities, pain, and disappointments of life. While my bad decisions were discussed on late-night talk shows, news programs, and magazines across America . . . it doesn't mean they're any different from anyone else's problems. If everyone had their "backstories" made by the producers of *Dancing with the Stars,* every single one would include dealing with life's challenges.

Again, I'm not a role model, a dancer, or a preacher. I'm just a normal girl who couldn't hide her problems and learned a few lessons along the way. Namely, that not being afraid of life's imperfection and complexity is the first step toward truly living it. Oh, and it helps to reach out to the only one who truly offers hope in this world.

No, not President Obama.

Isaiah 41:10 says, "Fear not, for I am with you; be not dismayed, for I am your God; I will strengthen you, I will help you, I will uphold you with my righteous right hand."

If you try to follow God's guidance for your life, he'll help you navigate around some of the big obstacles. But be warned, there's no telling where you might end up! He has a way of surprising you, or pushing you further than you think you can go.

Who knows? You might even end up in a gorilla suit of your own.

Acknowledgments

Tripp Easton—you are the light of my life, thank you for being such a good boy.

Mom and Dad, thank you for tough love.

Track, thank you for keeping me grounded.

Willow, thank you for constant entertainment and always giving me a listening ear.

Piper, thank you for having that little smile that always brightens my day.

Trig, thank you for showing us what's important in life.

Love you guys.

Dr. Cusack, Kelly, Kim, and Jenny—thank you for giving me opportunities when I was trying to make it on my own. I'll always be your "Susie," Doc!

Aunt Heather, thank you for taking me in, and for the best home cooking.

Aunt Molly, thank you for always supporting me, and introducing me to mascara.

To my *Dancing with the Stars* pals, Deena, Mark, and Mark's family, Claire, Florence, Kyle and his family . . . you made the experience so rewarding, and your friendship is much better than the disco ball.

To Karen, for being so loving and patient with my boy, and for watching the baseball movie a million times!

Grandma Sally, Nana, Marissa, Kandice, Lauden, McKinley, Britta, Jenna, Elle, Alex, Miranda, Janice, Crystal, Marina, Breann, Kelsie, Alexa, Juanita, Barb, Kate, Peg, Christie, Brandi, Kelly, Faye, thank you for making the good times more fun, and the bad times more bearable.

Grandpa Chuck, Papa Jim, Papa Bob, Payton, Heath, Karch, Landon, Kier, Teko, Nerd, Kurt, Jack, Jacob, The Jones, Jake, Ry, you boys are the best. Thank you for always having my back.

TVF, JJT, and Chelsea, you guys will always be the best team.

To Paige Adams Geller, the Wasilla girl who founded Paige Premium Denim and who keeps my entire family looking good in PPD!

To all the folks at HarperCollins, my editor, Amy Bendell, publicist Shelby Meizlik, and everyone who helped along the way.

To my wonderful writer, Nancy, who guided me through this process and captured my story with so much heart.

And lastly—to Hunter Wolfe, thank you for the flowers and cute little notes on my car. I will forever miss your cute, shy demeanor. Your life and death make me realize the importance of being honest about my own struggles, and of talking about the true hope that things, ultimately, can be better.